Survival Guide

for

General Chemistry
with Math Review

Charles H. Atwood
University of Georgia

THOMSON

BROOKS/COLE

Australia • Canada • Mexico • Singapore • Spain • United Kingdom • United States

Printed in the United States of America
5 6 7 07 06 05

Printer: Thomson/West

Cover Image: ©Art Valero/Images.com, Inc.

ISBN: 0-534-99370-2

For more information about our products,
contact us at:
Thomson Learning Academic Resource Center
1-800-423-0563

For permission to use material from this text,
contact us by:
Phone: 1-800-730-2214
Fax: 1-800-730-2215
Web: http://www.thomsonrights.com

Thomson Brooks/Cole
10 Davis Drive
Belmont, CA 94002-3098
USA

Asia
Thomson Learning
5 Shenton Way #01-01
UIC Building
Singapore 068808

Australia/New Zealand
Thomson Learning
102 Dodds Street
Southbank, Victoria 3006
Australia

Canada
Nelson
1120 Birchmount Road
Toronto, Ontario M1K 5G4
Canada

Europe/Middle East/South Africa
Thomson Learning
High Holborn House
50/51 Bedford Row
London WC1R 4LR
United Kingdom

Latin America
Thomson Learning
Seneca, 53
Colonia Polanco
11560 Mexico D.F.
Mexico

Spain/Portugal
Paraninfo
Calle/Magallanes, 25
28015 Madrid, Spain

Table of Contents

Module 5
Chemical Reaction Stoichiometry

Module 6
Types of Chemical Reactions

Module 7
Electronic Structure of Atoms

Module 8
Chemical Periodicity

Module 9
Chemical Bonding

Module 10
Molecular Shapes

Module 11
Hybridization and Polarity of Molecules

Module 12
Acids and Bases

Module 13
States of Matter

Module 14
Solutions

Module 15
Heat Transfer, Calorimetry, and Thermodynamics

Module 16
Chemical Kinetics

Module 17
Gas Phase Equilibria

Module 18
Aqueous Equilibria

Module 19
Electrochemistry

Module 20
Nuclear Chemistry

Math Review

Preface

To The Student:

Students in general chemistry are often overwhelmed with the amount of material that is covered in a typical course and with the confusing sets of rules that are required to be successful. In my position as an instructor of over 1500 students per year, I am often asked the same sets of questions about the topics encountered in my courses, just phrased differently for each student. In writing this Survival Guide, I have diligently tried to take those common questions and answer them as if you, the student, were sitting beside me in my office. The Guide is an attempt to let you see the necessary steps in the most frequently encountered problems from a typical General Chemistry course. The included Sample Exercises are typical test problems used here at the University of Georgia and which you may encounter in your course. To help you understand the thought processes needed for the problems I have included arrows that point out where the numbers and essential concepts are found. There are also **INSIGHT:** boxes that show you what to look for in these problems to help you determine the type of problem encountered. The included **YIELD** boxes are where I have tried to point out common mistakes, essential points that you must consider, and any mathematical issues that are essential to the correct solution of the problems. One of my main goals has been to condense the material to its bare essence so that you could focus on the major points without so much other material obstructing your view. In the development stages I have given many of the Modules to my students here at the University of Georgia and let them use the Modules. Their response has been overwhelmingly positive because they have found them to be the bare bones of the course. I hope that you will also have that experience and find this guide not only helps with your survival but helps you thrive in your course.

Acknowledgements:

The initial idea for this Guide came from Michelle Julet at Thomson. In conjunction with David Harris and Karoliina Tuovinen, we have taken Michelle's idea and either improved it or mangled it beyond her initial conception. In either case, I am most indebted to the hard work and support that all three of my colleagues at Brooks/Cole-Thomson have provided as I trudged through the writing of this Guide. In particular, David had the great ideas of the arrows, INSIGHT: and YIELD boxes. Dr. Lenore Polo Rodicio from Miami-Dade College was the reviewer for this work providing some most insightful suggestions and caught numerous errors. I appreciate her efforts.

Writing is a solitary occupation but one that I could not do without the support, guidance, and help of my loving family, Judy, Louis, and Lesley. The three of you make this work so much more fun. I thank you from the bottom of my heart.

I would like to dedicate this work to two other people that have been important in my life, my aunt O'Neita Dongieux and my sister, Peggy Jones. It seems appropriate that I dedicate this Survival Guide to the two of you because I would not have survived my youth without you both.

Module 1
Metric System, Significant Figures, Dimensional Analysis, and Density

Introduction

This module introduces the <u>basic rules of the metric system and significant figures, and how to use dimensional analysis to help solve problems, including density related problems.</u> These are typical topics introduced in the first chapter of general chemistry textbooks.

Module 1 Key Equations & Concepts

1. $d = m/V$

    The density equation is used to determine:

    (a) density, when mass and volume are given

    (b) mass, when density and volume are given

    (c) volume, when density and mass are given

The metric system uses a series of multipliers to convert from one sized unit to another size. You must be very familiar with these prefixes and how to convert from one size unit to another. A common set of multiplier prefixes is given in this table.

Prefix Name	Prefix Symbol	Multiplication Factor
mega-	M	1000000 or 10^6
kilo-	k	1000 or 10^3
deci-	d	0.1 or 10^{-1}
centi-	c	0.01 or 10^{-2}
milli-	m	0.001 or 10^{-3}
micro-	μ	0.000001 or 10^{-6}
nano-	n	0.000000001 or 10^{-9}
pico-	p	0.000000000001 or 10^{-12}

Metric System Sample Exercises

1. *How many mm are there in 3.45 km?*

    The correct answer is 3.45×10^6 mm.

    The table indicates that there are 1000 m in 1 km and that 1 mm = 0.001 m.

    Converts km to m. Converts m to mm.

$$? \, \text{mm} = 3.45 \, \text{km} \left(\frac{1000 \, \text{m}}{1 \, \text{km}} \right) \left(\frac{1 \, \text{mm}}{0.001 \, \text{m}} \right) = 3.45 \times 10^6 \, \text{mm}$$

Notice that there are millions of mm in a km.

One way to help insure that you work these problems correctly is to remember which one of the units is the largest. In this problem, the km is a much larger unit than the mm. Thus we should expect that there will be many of the smaller unit, mm's, in the large units. Notice that the answer indicates there are 3.45 million mm in 3.45 km, which is sensible.

2. ***How many mg are there in 15.0 pg?***
   The correct answer is 1.5×10^{-8} mg.
   From the table we see that 1 pg = 10^{-12} g and 1 mg = 10^{-3} g.

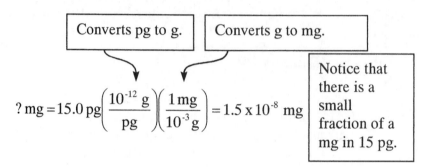

Converts pg to g.   Converts g to mg.

$$? \, mg = 15.0 \, pg \left(\frac{10^{-12} \, g}{pg} \right) \left(\frac{1 \, mg}{10^{-3} \, g} \right) = 1.5 \times 10^{-8} \, mg$$

Notice that there is a small fraction of a mg in 15 pg.

In this problem picograms, pg, are the smaller unit. We should expect that there are very few milligrams, mg, in 15.0 pg. Consequently, the answer 1.5×10^{-8} mg is reasonable.

Significant Figures Sample Exercises

3. ***How many significant figures are in the number 58062?***
   The correct answer is five significant figures.

This zero is significant because it is embedded in other significant digits.

4. ***How many significant figures are in the number 0.0000543?***
   The correct answer is three significant figures.

These zeroes are not significant because their purpose is to indicate the position of the decimal place.

5. ***How many significant figures are in the number 0.009120?***
   The correct answer is four significant figures.

These three zeroes are not significant just as in exercise 4.

This zero is significant.

Zeroes at the end of a number that includes a decimal point are significant.

6. ***How many significant figures are in the number 24500?***

These zeroes may or may not be significant.

In this problem it is unclear how many significant digits are present. There could be as few as three (the 2, 4, and 5) or as many as five if both of the zeroes are significant.

2

7. *How many significant figures are in the number 2.4500 x 10⁴?*

The correct answer is five significant figures.

$$2.4500 \times 10^4$$

> As written both of these zeroes are significant.

Notice that this number is the same as in exercise 6 but written in scientific notation. Furthermore, it uses the rule displayed in exercise 5 as well. Thus the significant digits are the 2, 4, 5, and both of the zeroes.

8. *What is the sum of 12.674 + 5.3150 + 486.9?*

The correct answer is 504.9.

> This 9 is in the tenths decimal place. It is the most doubtful digit in the sum.

In addition and subtraction problems involving significant figures the final answer must contain no digits beyond the most doubtful digit in the numbers being added or subtracted. The most doubtful digit in each of the numbers is underlined 12.674, 5.3150, 486.9. Notice that the 486.9 has the most doubtful digit because the 9 is only in the tenths position and the other numbers are doubtful in the thousandths (12.674) and ten thousandths (5.3150) positions. The final answer must have the final digit in the tenths position.

9. *What is the correct answer to this problem, 2.6138 x 10⁶ – 7.95 x 10⁻³?*

The correct answer is 2.6138×10^6.

> This 8 is the most doubtful digit in the sum. It is in the hundreds position.

The number 2.6138×10^6 can be also written as 2,613,800. Its most doubtful digit, the 8, is in the hundreds position. The other number, 7.98×10^{-3}, can be written as 0.00795. Its most doubtful digit, the 5, is in the one millionths position. Consequently, the final answer cannot extend beyond the 8 in 2.6138×10^6. When adding and subtracting, both numbers must be expressed to the same power of 10 in order to determine the most doubtful digit.

10. *What is the correct answer to this problem, 47.893 x 2.64?*

The correct answer is 126.

> This number contains only 3 significant digits. It determines the final result.

In multiplication and division problems involving significant figures the final answer must contain the same number of significant figures as the number with the least number of significant figures. In this problem 47.893 has five significant figures and 2.64 has three significant figures. The correct answer must have three significant figures to match the number of significant figures in 2.64, thus the answer is 126.

11. *What is the correct answer to this problem, 1.95x10⁵÷7.643 x 10⁻⁴?*

The correct answer is 2.55×10^8.

> This number contains 3 significant figures.

> This number contains 4 significant figures.

Just as in exercise 10, the number with fewest significant digits, 1.95×10^5 having three significant digits, determines the final answer of 2.55×10^8 also having three significant digits.

Dimensional Analysis Sample Exercises

In chemistry we often perform calculations that require changing from one set of units, say ft or in^2 or cm^3, to a second set of units like Mm or km^2 or yd^3. Dimensional analysis is a convenient method to help convert units without making arithmetic errors. In this method common conversion factors, which are given in your textbook, are arranged so that one set of units cancels converting the problem to the second set of units.

12. How many Mm are in 653 ft?

The correct answer is 1.99×10^{-4} Mm.

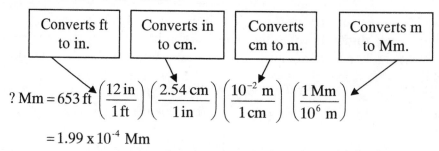

$$? \, Mm = 653 \, ft \left(\frac{12 \, in}{1 \, ft} \right) \left(\frac{2.54 \, cm}{1 \, in} \right) \left(\frac{10^{-2} \, m}{1 \, cm} \right) \left(\frac{1 \, Mm}{10^6 \, m} \right)$$

$$= 1.99 \times 10^{-4} \, Mm$$

Notice that the problem is arranged so that each successive conversion factor makes progress in the conversion process. Feet are converted to inches, then to cm, next to m, and finally to Mm. This is the simplest kind of dimensional analysis problem because all of the units are linear. The next problem illustrates a conversion problem involving two dimensional units.

13. How many km^2 are in 2.5×10^8 in^2?

The correct answer is 1.6×10^{-1} km^2.

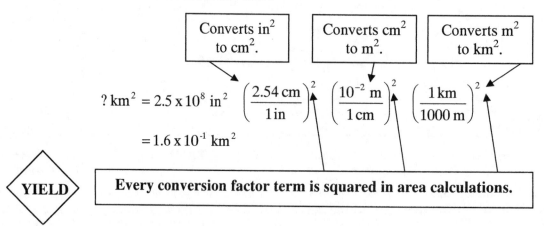

$$? \, km^2 = 2.5 \times 10^8 \, in^2 \left(\frac{2.54 \, cm}{1 \, in} \right)^2 \left(\frac{10^{-2} \, m}{1 \, cm} \right)^2 \left(\frac{1 \, km}{1000 \, m} \right)^2$$

$$= 1.6 \times 10^{-1} \, km^2$$

⬦ YIELD ⬦ **Every conversion factor term is squared in area calculations.**

Because the problem involves area, which is a two dimensional unit, all of the conversion factors are similar to exercise 12 but must be squared to be in the appropriate units.

14. How many yd^3 are in 7.93×10^{12} cm^3?

The correct answer is 1.04×10^7 yd^3.

4

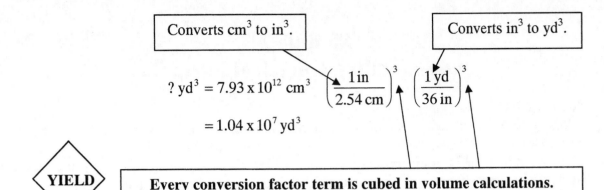

Converts cm³ to in³.   Converts in³ to yd³.

$$? \, yd^3 = 7.93 \times 10^{12} \, cm^3 \left(\frac{1 \, in}{2.54 \, cm}\right)^3 \left(\frac{1 \, yd}{36 \, in}\right)^3$$

$$= 1.04 \times 10^7 \, yd^3$$

⟨YIELD⟩ | **Every conversion factor term is cubed in volume calculations.**

In this problem all of the conversion factors are cubed because volume is involved.

Density Sample Exercises

15. *What is the mass, in g, of a 68.2 cm³ sample of ethyl alcohol? The density of ethyl alcohol is 0.789 g/cm³.*
    The correct answer is 53.8 g.

The density converts the volume of a substance into the mass.

$$? \, g = 68.2 \, cm^3 \left(\frac{0.789 \, g}{1 \, cm^3}\right)$$

$$= 53.8 \, g$$

Final units are g because the cm³ in the density cancels with the original volume.

16. *What is the volume, in cm³, of a 237.0 g sample of copper? The density of copper is 8.92 g/cm³.*
    The correct answer is 26.6 cm³.

Density, inverted, cancels the original mass, in g, leaving the volume in cm³.

$$? \, cm^3 = 237.0 \, g \left(\frac{1 \, cm^3}{8.92 \, g}\right)$$

$$= 26.6 \, cm^3$$

17. *What is the density of a substance having a mass of 25.6 g and a volume of 74.3 cm³?*
    The correct answer is 0.345 g/cm³.

$$? \, g/cm^3 = \frac{25.6 \, g}{74.3 \, cm^3} = 0.345 \, g/cm^3$$

Density's units, g/cm³, help determine the correct order of division.

Module 2
Understanding Chemical Formulas

Introduction

What information is contained in a chemical formula and how do we interpret that information? Chemists use specific symbolism to express their understanding of elements, compounds, ions and ionic compounds. In this module we will look at these symbols and learn how to recognize the number and types of atoms or ions present.

Module 2 Key Equations & Concepts

1. C_5H_{12}
   Molecular formulas indicate the number of each atom present in a molecule.
2. $Al_2(CO_3)_3$
   Ionic formulas indicate the number of each ion present in a formula unit.
3. $4\ C_5H_{12}$
   Stoichiometric coefficients indicate the number, usually in moles, of a particular molecule or formula unit in the chemical symbolism.

Interpreting Chemical Formulas Sample Exercises

1. *How many atoms of each element are present in one molecule of C_2H_5OH?*
   The correct answer is 2 carbon atoms, 1 oxygen atom, and 6 hydrogen atoms.

$$C_2H_5OH$$

   There are 2 carbon atoms, 1 oxygen atom, and 6 hydrogen atoms in one molecule.
   The molecular formula displays the number of atoms in each molecule of a species.

2. *How many atoms of each element are present in one formula unit of $Al_2(SO_4)_3$?*
   The correct answer is 2 aluminum atoms, 3 sulfur atoms, and 12 oxygen atoms.

$$Al_2(SO_4)_3$$

   There are 2 aluminum atoms, 3 sulfur atoms, and 12 oxygen atoms.
   Remember, numbers outside a parenthesis are multiplied times the subscripts of all the elements inside the parentheses. Thus there are 3 x 1 = 3 sulfur atoms and 3 x 4 = 12 oxygen atoms.

Using Stoichiometric Coefficients Sample Exercises

3. *How many atoms of each element are present in this chemical formula, $3\ C_5H_{12}$?*
   The correct answer is 15 C atoms and 36 hydrogen atoms.

$$3\ C_5H_{12}$$

There are 15 carbon atoms and 36 hydrogen atoms in $3\ C_5H_{12}$.

$3 \times 5 = 15$ C atoms and $3 \times 12 = 36$ H atoms

4. ***How many atoms of each element are present in this chemical formula,***
$5\ Ca_3(PO_4)_2$?

The correct answer is 15 calcium atoms, 10 phosphorus atoms, and 40 oxygen atoms.

$$5\ Ca_3(PO_4)_2$$

There are 15 calcium atoms, 10 phosphorus atoms, and 40 oxygen atoms.

$5 \times 3 = 15$ Ca atoms, $5 \times 2 = 10$ P atoms, and $5 \times 4 \times 2 = 40$ O atoms

Interpreting Chemical Formulas Sample Exercises

5. ***Using circles to represent the atoms, draw your best representation of what***
C_4H_{10} looks like, if we could see atoms, ions, and molecules.

$$C_4H_{10}$$

The 4 carbon atoms are in the center of the molecule.

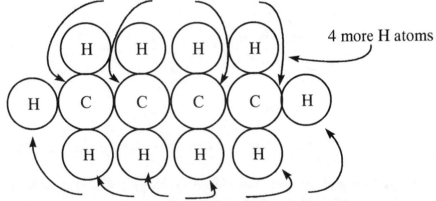

The 10 H atoms are around the outside of the molecule.
Notice, that this is one single molecule not 14 separate things.

From a chemical standpoint this is not the only way to draw C_4H_{10} but all of the possibilities will consist of molecules with the atoms connected.

6. Using circles to represent the atoms and ions, draw your best representation of what Sr₃(PO₄)₂ looks like, if we could see atoms, ions, and molecules. Remember, ions are independent species unto themselves.

Notice that the three Sr^{2+} ions are independent species.

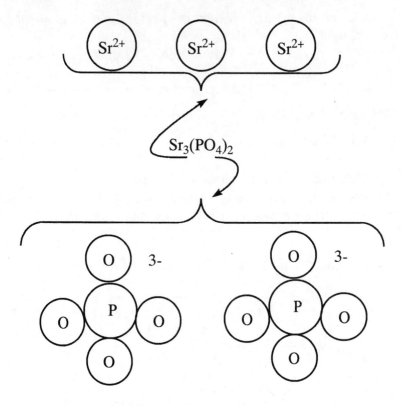

The two PO_4^{3-} ions are also independent species.

All of the formulas and symbols introduced up to now can also be used to represent moles of a species. Thus, if asked how many atoms, ions, or molecules there are in one mole of each of these species, simply multiply the answers given above by Avogadro's number, 6.022×10^{23}. Remember that 1 mole = 6.022×10^{23}, just like 1 dozen = 12. This next problem illustrates this idea.

Using Chemical Formulas to Determine Numbers of Atoms in One Mole of a Substance Sample Exercise

  7. How many atoms of each element are present in one mole of Al₂(SO₄)₃?

$$Al_2(SO_4)_3$$

There are:
2 x 6.022×10^{23} = 12.044×10^{23} aluminum atoms,
3 x 6.022×10^{23} = 18.066×10^{23} sulfur atoms, and
12 x 6.022×10^{23} = 72.264×10^{23} oxygen atoms.

Module 3
Chemical Nomenclature

Introduction

In this module we will examine the naming of a variety of simple inorganic chemical compounds. The key to these exercises is <u>recognition of the type of chemical compound and application of the appropriate nomenclature rules.</u>

Module 3 Key Equations & Concepts

1. **Metal cations combined with nonmetal anions.**
   Simple binary ionic compounds - nomenclature is metal's name followed by nonmetal's stem plus –ide.

2. **Metal cations combined with polyatomic anions.**
   Pseudobinary ionic compounds – nomenclature is metal's name followed by polyatomic ion's name.

3. **Transition metal cations combined with nonmetal anions.**
   Transition metal ionic compounds – nomenclature is metal's name with oxidation state in parentheses followed by nonmetal's stem plus –ide.

4. **Two nonmetals combined in one compound.**
   Binary covalent compounds – less electronegative element named first, more electronegative named second using stem plus –ide. Prefixes such as di-, tri-, etc. are used for both elements.

5. **Hydrogen combined with a nonmetal in aqueous solution.**
   Binary acids – nomenclature is hydro followed by nonmetal stem with suffix –ic acid.

6. **Hydrogen, oxygen, and a nonmetal combined in one compound.**
   Ternary acids – nomenclature is a series of names based upon the oxidation state of the nonmetal.
   Nonmetal highest oxidation state is per stem –ic acid.
   Nonmetal second highest oxidation state is stem –ic acid.
   Nonmetal third highest oxidation state is stem –ous acid.
   Nonmetal lowest oxidation state is hypo stem –ous acid.

7. **Metal ions combined with a polyatomic ion made from a ternary acid.**
   Ternary acid salts – nomenclature is the metal's name followed by the same series of names used for the ternary acid with two changes. The –ic suffixes are changed to –ate and the –ous suffixes are changed to –ite.

8. **Metal ions combined with a ternary acid salt and hydrogen.**
   Acidic salts of ternary acids – nomenclature is the metal's name followed by hydrogen (including the appropriate di-, tri-, etc. prefix) plus the ternary acid salt's name.

9. **Metal ions combined with hydroxyl groups and nonmetal ions.**
   Basic salts of polyhydroxy bases – nomenclature is the metal's name followed by hydroxy (including the appropriate di-, tri-, etc. prefix) plus the nonmetal stem plus –ide.

Chemical Nomenclature Sample Exercises

  1. What is the correct name of this chemical compound, CaBr$_2$?

  The correct name is *calcium bromide*.

  Ca^{2+} is a positive ion made from a metal.

  Br^{1-} ions are negative ions made from the nonmetal bromine.

Metal cations and nonmetal anions make simple binary ionic compounds. Simple binary ionic compounds are named using the metal's name followed by the nonmetal's stem and the suffix –ide. No prefixes like di- or tri- are used to denote the number of ions present in the substance.

  2. What is the correct name of this chemical compound, Mg$_3$(PO$_4$)$_2$?

  The correct name is *magnesium phosphate*.

  Mg^{2+} is a positive ion made from a metal.

  PO_4^{3-} is a negative polyatomic ion made from phosphoric acid.

Metal cations and polyatomic anions make pseudobinary ionic compounds. These compounds are named using the metal's name followed by the correct name of the polyatomic anion. Your textbook has a list of the polyatomic anions that you are expected to know. Make sure that you have the name, the anion's formula, and the charge memorized. Once again, no prefixes are used in these compounds to tell the number of ions present.

  3. What is the correct name of this chemical compound, FeCl$_3$?

  The correct name is iron(III) chloride.

  Fe^{3+} is a positive ion made from a transition metal (B Groups on the periodic chart).

  Cl^{1-} is a negative ion made from a nonmetal.

Transition metal cations and nonmetal or polyatomic anions make transition metal ionic compounds. Their names are derived from the metal's name followed by the metal's oxidation state in Roman numerals inside parentheses. A metal's oxidation state is determined from the oxidation state of the anion. Notice that in all of the ionic compounds up to this point, prefixes have not been used to tell the number of ions present.

  4. What is the correct name of this chemical compound, N$_2$O$_4$?

  The correct name is dinitrogen tetroxide.

  This compound is made from two nonmetals, nitrogen and oxygen.

Two nonmetals form binary covalent compounds. These compounds use prefixes to indicate the number of atoms of each element present in the compound. This is an important difference from the ionic compounds used before.

  5. What is the correct name of this chemical compound, H$_2$S(aq)?

  The correct name is hydrosulfuric acid.

  This compound is made from hydrogen and a nonmetal.

Furthermore, the symbol (aq) also indicates that this compound is dissolved in water. That combination is indicative of a binary acid. Binary acids are named using the prefix hydro- followed by the nonmetal's stem and the suffix –ide. Be careful. If the symbol (aq) is not present, then the compound is named as a binary covalent compound. In this case H$_2$S without the (aq) would be named dihydrogen sulfide.

  6. What is the correct name of this chemical compound, HIO$_3$?

  The correct name is iodic acid.

> This compound is made from three nonmetals, H, O, and another nonmetal, iodine.

This combination of nonmetals is a <u>ternary acid</u>. Ternary acids are named based on a system derived from the third nonmetal's oxidation state. The easiest method to learn these compounds is to use the table of "ic acids" found in your textbook. You must learn both the compound's formula and its name. Once you know the "ic acids" then use the following system. The acid with one more O atom than the "ic acid" is the "per stem ic acid". One fewer O atom than the "ic acid" is the "ous acid". Two fewer O atoms than the "ic acid" is the "hypo stem ous acid". Here is the entire series for the iodine ternary acids.

HIO_4 is periodic acid. Contains one more O atom than HIO_3 and the I atom has an oxidation state of +7. HIO_3 is iodic acid with an oxidation state of +5 for I. HIO_2 is iodous acid containing an I atom with oxidation state +3. Finally, HIO is hypoiodous acid. The I atom is in the +1 oxidation state. You will be expected to know all of these various acids.

7. ***What is the correct name of this chemical compound, KIO_4?***
> The correct name is potassium periodate.
> This compound is made from a metal ion, K^+, and a polyatomic anion that is derived from the ternary acids discussed above.

This is called a <u>ternary acid salt</u>. The anion's name is based upon the ending of the ternary acid. Ternary acids ending in "ic" give salts that end in "ate". Ternary acids that end in "ous" give salts that end in "ite". The prefixes per- and hypo- are retained. The iodic acid series of potassium salts are shown below.

KIO_4 is potassium periodate. KIO_3 is potassium iodate. KIO_2 is potassium iodite. Finally, KIO is potassium hypoiodite. These are some of the most difficult compounds to name. Work hard on these.

8. ***What is the correct name of this chemical compound, NaH_2PO4?***
> The correct name is sodium dihydrogen phosphate.
> This compound is made from a metal cation, two Na^+ ions, and a polyatomic anion made from a ternary acid that still retains some of its acidic hydrogens.

These compounds are called <u>acidic salts of ternary acids</u>. The names for these compounds use the word hydrogen plus a prefix, in this case di-, to indicate the number of acidic hydrogens that are present. The last part of the salt's name is the same as determined in question 7 for the ternary acid salts.

9. ***What is the correct name of this chemical compound, $Al(OH)_2Cl$?***
> The correct name is aluminum dihydroxy chloride.
> This compound is made from a metal ion, Al^{3+}, and three anions two of which are hydroxide ions.

Compounds containing hydroxide ions and other anions plus a metal ion are <u>basic salts of polyhydroxy bases</u>. Their name must indicate the number of OH^{1-} groups that are present in the compound. This is done using the appropriate prefix attached to hydroxy. The remainder of the compound's name is the same as for binary ionic compounds. See question 1 above for those rules.

Module 4
The Mole Concept

Introduction

In this module we will examine several equations that are used in problems involving the mole concept. The important thing to understand from this module is how to interpret, use, and perform all of the important calculations that involve the mole. You will need a periodic table as you work all of these exercises. Atomic weights come directly from the periodic table.

Module 4 Key Equations & Concepts

1. **Molar mass** $= \sum$ **atomic weights of atoms in a molecule or ion**

   The molar mass, molecular weight, or formula weight equation is used to determine the mass in grams of one mole of a substance.

2. **1 mole = 6.022 x 10^{23} particles**

   Avogadro's relationship is used to convert from the number of moles of a substance to the number of atoms, ions, or molecules of a substance and vice versa.

3. **mass of one atom of an element** $= \left(\dfrac{\text{grams of an element}}{\text{1 mole of an element}} \right) \left(\dfrac{\text{1 mole of atoms}}{6.022 \times 10^{23} \text{ atoms}} \right)$

   The mass of one atom, ion, or molecule is used to determine the mass of a few atoms, ions, or molecules of a substance.

Determining the Molar Mass Sample Exercise

1. What is the molar mass or formula weight of calcium phosphate, Ca₃(PO₄)₂?

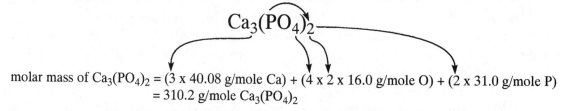

molar mass of $Ca_3(PO_4)_2$ = (3 x 40.08 g/mole Ca) + (4 x 2 x 16.0 g/mole O) + (2 x 31.0 g/mole P)
= 310.2 g/mole $Ca_3(PO_4)_2$

Determining the Number of Moles Sample Exercise

2. How many moles of calcium phosphate are there in 65.3 g of Ca₃(PO₄)₂?

$$? \text{ moles of } Ca_3(PO_4)_2 = 65.3 \text{ g } Ca_3(PO_4)_2 \left(\frac{1 \text{ mol } Ca_3(PO_4)_2}{310.2 \text{ g } Ca_3(PO_4)_2} \right)$$

$$= 0.211 \text{ mol } Ca_3(PO_4)_2$$

Molar mass of calcium phosphate from exercise #1.

Once the number of moles of the sample is known, we can determine the number of molecules or formula units of the substance. (Molecules are found in covalent compounds. Ionic compounds do not have molecules thus their smallest subunits are named formula units.)

Determining the Number of Molecules or Formula Units Sample Exercise
> *3. How many formula units of calcium phosphate are there in 0.211 moles of Ca$_3$(PO$_4$)$_2$?*

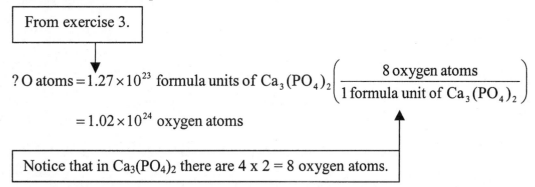

$$? \text{ formula units of } Ca_3(PO_4)_2 = 0.211 \text{ moles of } Ca_3(PO_4)_2 \left(\frac{6.022 \times 10^{23} \text{ formula units}}{1 \text{ mole of } Ca_3(PO_4)_2} \right)$$

$$= 1.27 \times 10^{23} \text{ formula units of } Ca_3(PO_4)_2$$

Determining the Number of Atoms or Ions Sample Exercise
> *4. How many oxygen, O, atoms are there in 0.211 moles of Ca$_3$(PO$_4$)$_2$?*

Using the last equation from this module, we can determine the mass of a few molecules or formula units of a compound.

From exercise 3.

$$? \text{ O atoms} = 1.27 \times 10^{23} \text{ formula units of } Ca_3(PO_4)_2 \left(\frac{8 \text{ oxygen atoms}}{1 \text{ formula unit of } Ca_3(PO_4)_2} \right)$$

$$= 1.02 \times 10^{24} \text{ oxygen atoms}$$

Notice that in Ca$_3$(PO$_4$)$_2$ there are 4 x 2 = 8 oxygen atoms.

Determining the Mass of Small Numbers of Molecules or Formula Units of a Substance Sample Exercise
> *5. What is the mass, in grams, of 25.0 formula units of Ca$_3$(PO$_4$)$_2$?*

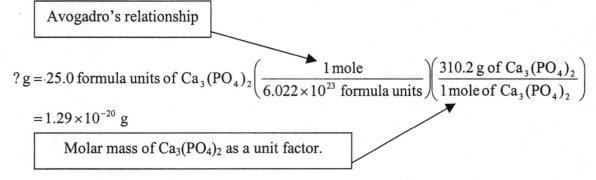

Avogadro's relationship

$$? \text{ g} = 25.0 \text{ formula units of } Ca_3(PO_4)_2 \left(\frac{1 \text{ mole}}{6.022 \times 10^{23} \text{ formula units}} \right) \left(\frac{310.2 \text{ g of } Ca_3(PO_4)_2}{1 \text{ mole of } Ca_3(PO_4)_2} \right)$$

$$= 1.29 \times 10^{-20} \text{ g}$$

Molar mass of Ca$_3$(PO$_4$)$_2$ as a unit factor.

13

Combined Equations Sample Exercise

6. How many carbon, C, atoms are there in 0.375 g of $C_4H_8O_2$?

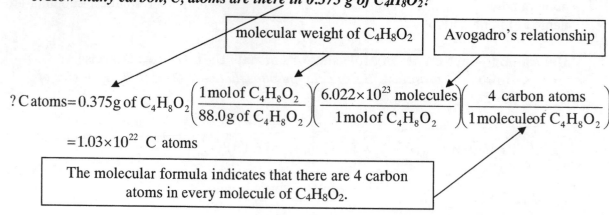

molecular weight of $C_4H_8O_2$

Avogadro's relationship

$$? \, C \, atoms = 0.375 \, g \, of \, C_4H_8O_2 \left(\frac{1 \, mol \, of \, C_4H_8O_2}{88.0 \, g \, of \, C_4H_8O_2} \right) \left(\frac{6.022 \times 10^{23} \, molecules}{1 \, mol \, of \, C_4H_8O_2} \right) \left(\frac{4 \, carbon \, atoms}{1 \, molecule \, of \, C_4H_8O_2} \right)$$

$$= 1.03 \times 10^{22} \, C \, atoms$$

The molecular formula indicates that there are 4 carbon atoms in every molecule of $C_4H_8O_2$.

Module 5
Chemical Reaction Stoichiometry

Introduction

In this module we will look at several problems that involve reaction stoichiometry. The important points to learn in this module are <u>balancing chemical reactions, basic reaction stoichiometry, limiting reactant calculations, percent yield calculations, and reactions in solution</u>. You will need a periodic table to calculate molecular weights in these problems.

Module 5 Key Equations & Concepts

1. **Percent yield = (actual yield ÷ theoretical yield) x 100**
   The percent yield formula is used to determine the percentage of the theoretical yield that was formed in a reaction.
2. **M x L = moles or M x mL = mmol**
   The relationship of molarity and volume is used to convert from solution concentrations to moles or from volume of a solution to moles of a solution.

Chemical reactions symbolically display what happens when chemical substances are mixed and new substances are formed. There is a certain amount of terminology that is required in this section.

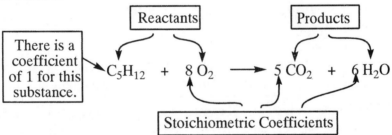

Stoichiometric coefficients are required to "balance" the equation. This process insures that equal numbers of atoms of each element are present in both the products and reactants. Otherwise the reaction would violate the Law of Conservation of Mass indicating that new atoms have been made in the reaction rather than new chemical compounds.

Balancing Chemical Reaction Exercises

1. Balance this chemical reaction using the smallest whole numbers.

$$Ca(OH)_2 + H_3PO_4 \rightarrow Ca_3(PO_4)_2 + H_2O$$

The correct answer is:

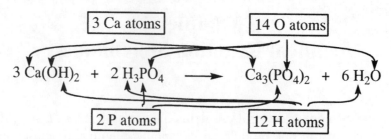

2. Balance this chemical reaction using the smallest whole numbers.

$$C_6H_{14} + O_2 \rightarrow CO_2 + H_2O$$

The correct answer is:

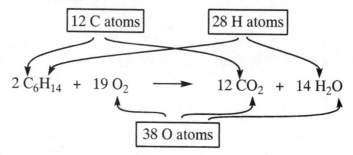

Simple Reaction Stoichiometry Exercises

3. How many moles of H_2 can be formed from the reaction of 3.0 moles of Na with excess H_2O.

$$2\,Na + 2\,H_2O \rightarrow 2\,NaOH + H_2$$

The correct answer is 1.5 moles of H_2.

INSIGHT:	The word **excess** is important. It is your clue that this problem does **not** involve a limiting reactant calculation.
	The reaction ratio, which comes from the balanced reaction, is 2 moles of Na consumed for every 1 mole of H_2 formed.

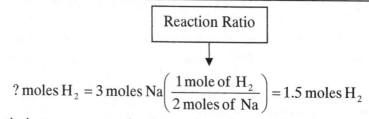

$$? \text{ moles } H_2 = 3 \text{ moles Na}\left(\frac{1 \text{ mole of } H_2}{2 \text{ moles of Na}}\right) = 1.5 \text{ moles } H_2$$

The reaction ratio is a new conversion factor that relates moles of any reactant or product to moles of another reactant or product.

4. How many grams of H_2 can be formed from the reaction of 11.2 grams of Na with excess H_2O?

$$2\,Na + 2\,H_2O \rightarrow 2\,NaOH + H_2$$

The correct answer is 0.494 g of H_2.

Converts g of Na to moles of Na	Reaction Ratio	Converts moles of H_2 to g of H_2

$$? \text{ grams of } H_2 = 11.2 \text{ g} \left(\frac{1 \text{ mole of Na}}{22.9 \text{ g of Na}} \right) \left(\frac{1 \text{ mole of } H_2}{2 \text{ moles of Na}} \right) \left(\frac{2.02 \text{ g of } H_2}{1 \text{ mole of } H_2} \right) = 0.494 \text{ g of } H_2$$

This problem makes the complete transformation from grams of one of the reactants, Na, to grams of one of the products, H_2. There is a very common set of transformations that are used in this calculation which will be used in many reaction stoichiometry problems.

grams of X $\rightarrow$ moles of X $\rightarrow$ reaction ratio $\rightarrow$ moles of Y $\rightarrow$ grams of Y

Limiting Reactant Sample Exercise

5. ***What is the maximum number of grams of H_2 that can be formed from the reaction of 11.2 grams of Na with 9.00 grams of H_2O?***

$$2 \, Na + 2 \, H_2O \rightarrow 2 \, NaOH + H_2$$

The correct answer is 0.494 g of H_2.

INSIGHT: The word excess is not in this problem and amounts of both reactants are given. These are your *clues that this is a limiting reactant problem*.

$$? \text{ grams of } H_2 = 11.2 \text{ g of Na} \left(\frac{1 \text{ mole of Na}}{22.9 \text{ g of Na}} \right) \left(\frac{1 \text{ mole of } H_2}{2 \text{ moles of Na}} \right) \left(\frac{2.02 \text{ g of } H_2}{1 \text{ mole of } H_2} \right) = 0.494 \text{ g of } H_2$$

$$? \text{ grams of } H_2 = 9.00 \text{ g of } H_2O \left(\frac{1 \text{ mole of } H_2O}{18.0 \text{ g of } H_2O} \right) \left(\frac{1 \text{ mole of } H_2}{2 \text{ moles of } H_2O} \right) \left(\frac{2.02 \text{ g of } H_2}{1 \text{ mole of } H_2} \right) = 0.505 \text{ g of } H_2$$

You must perform reaction stoichiometry steps for every reactant for which an amount was given in the problem. In this case that is two steps.

YIELD **The maximum amount will be the _smallest_ amount that you calculate in the reaction stoichiometry steps!**

Percent Yield Sample Exercise

6. ***If 11.2 g of Na reacts with 9.00 g of H_2O and 0.400 g of H_2 is formed, what is the percent yield of the reaction?***

$$2 \, Na + 2 \, H_2O \rightarrow 2 \, NaOH + H_2$$

This is the actual yield.

The correct answer is 81.0%.

INSIGHT: Key clues that indicate percent yield problems are a) amounts of both reactants, b) an amount for the product, and c) the words percent yield.

In percent yield problems limiting reactant calculations are frequently performed first to determine the theoretical yield. For this problem the theoretical yield is the same as determined in exercise 5, 0.494 g of H_2.

$$\% \text{ yield} = \frac{\text{actual yield}}{\text{theoretical yield}} \times 100 = \frac{0.400 \text{ g}}{0.494 \text{ g}} \times 100\% = 81.0\%$$

Reactions in Solution Sample Exercise

7. **How many mL of 0.250 M HCl are required to react with 15.0 mL of 0.150 M Ba(OH)₂?**

$$2 \, HCl_{(aq)} + Ba(OH)_{2(aq)} \rightarrow BaCl_{2(aq)} + 2 \, H_2O_{(\ell)}$$

The correct answer is 18.0 mL of 0.250 M HCl.

INSIGHT: Key clues to indicate a reaction in solution are the presence of solutions (0.250 M) and mL in the problem.

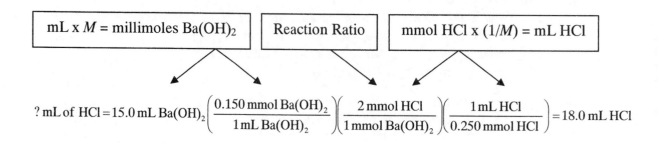

Module 6
Types of Chemical Reactions

Introduction

This module focuses on recognizing several of the simplest types of chemical reactions and predicting the products of these reactions. The most important points to learn in this module are <u>how to use the reactants of a chemical reaction to discern the type of reaction, predict the reaction products of metathesis reactions, plus how to write the total and net ionic equations for reactions.</u>

Module 6 Key Equations & Concepts

1. $KOH_{(aq)} + HI_{(aq)} \rightarrow KI_{(aq)} + H_2O_{(\ell)}$ – **Formula Unit Equation**
   The formula unit equation shows all of the species involved in a reaction as ionic or molecular compounds.

2. $K^+_{(aq)} + OH^-_{(aq)} + H^+_{(aq)} + I^-_{(aq)} \rightarrow K^+_{(aq)} + I^-_{(aq)} + H_2O_{(\ell)}$ – **Total Ionic Equation**
   The total ionic equation shows all of the ions in their ionized states in solution. All species that ionize completely in water are shown as separated ions.

3. $OH^-_{(aq)} + H^+_{(aq)} \rightarrow H_2O_{(\ell)}$ – **Net Ionic Equation**
   To write the net ionic equation, remove all spectator ions from the total ionic equation. Spectator ions are species that do not change as the reaction proceeds from reactants to products.

Reduction Oxidation Reaction Sample Exercise

1. What reaction types are represented by this chemical reaction?

$$2\,Na + 2\,H_2O \rightarrow 2\,NaOH + H_2$$

The correct answer is reduction oxidation (redox) and combination reaction.

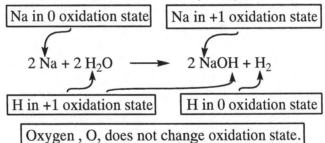

Na in 0 oxidation state	Na in +1 oxidation state

$$2\,Na + 2\,H_2O \longrightarrow 2\,NaOH + H_2$$

H in +1 oxidation state	H in 0 oxidation state

Oxygen, O, does not change oxidation state.

INSIGHT: To recognize redox reactions you must *look for chemical species that are changing their oxidation states.*

Your textbook has a series of rules for assigning oxidation numbers to elements in chemical species. If you do not know the rules for oxidation states, learn them now.

Combination Reaction Sample Exercise

2. *What reaction types are represented by this chemical reaction?*

$$3\,Sr\ +\ N_2\ \rightarrow\ Sr_3N_2$$

The correct answer is combination and reduction oxidation (redox) reactions.

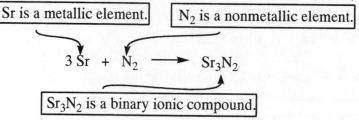

| Sr is a metallic element. | | N₂ is a nonmetallic element. |

$$3\,\dot{S}r\ +\ N_2\ \longrightarrow\ Sr_3N_2$$

Sr₃N₂ is a binary ionic compound.

INSIGHT: Combination reactions are characterized by *a) the reaction of two elements to form a compound, b) the reaction of a compound and an element to form a new compound, or c) the reaction of two compounds to form a new element.*

Combination reactions may also frequently be classified as another reaction type. In this case the second classification is a redox reaction. Sr is oxidized from the 0 to the +2 oxidation state and N_2 is reduced from the 0 to the -3 oxidation state.

Decomposition Reaction Sample Exercise

3. *What reaction types are represented by this chemical reaction?*

$$2\,CaO\ \rightarrow\ 2\,Ca\ +\ O_2$$

The correct answer is decomposition and reduction oxidation (redox) reactions.

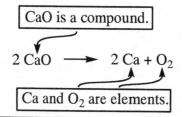

CaO is a compound.

$$2\,CaO\ \longrightarrow\ 2\,Ca + O_2$$

Ca and O₂ are elements.

INSIGHT: There are three types of decomposition reactions: *a) compounds decomposing into two or more elements, b) compounds decomposing into another compound and an element, and c) compounds decomposing into two simpler compounds.*

Decomposition reactions are the reverse of combination reactions. Instead of putting elements or compounds together to make new compounds decomposition reactions break compounds into elements or less complex compounds. As in combination reactions, decomposition reactions can frequently be classified as other reaction types. In this case the Ca^{2+} ion is reduced to Ca in the 0 oxidation state and the O^{2-} ion is oxidized to the 0 oxidation state in O_2.

Displacement Reaction Sample Exercise

4. *What reaction types are represented by this reaction?*

$$2\,Al\ +\ 3\,H_2SO_4\ \rightarrow\ Al_2(SO_4)_3\ +\ 3\,H_2$$

The correct answer is displacement and reduction oxidation (redox) reactions.

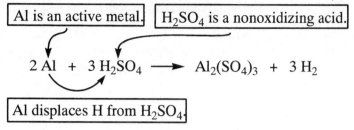

| Al is an active metal. | | H$_2$SO$_4$ is a nonoxidizing acid. |

$$2\,Al \; + \; 3\,H_2SO_4 \longrightarrow Al_2(SO_4)_3 \; + \; 3\,H_2$$

Al displaces H from H$_2$SO$_4$.

INSIGHT: Displacement reactions are characterized by *one element replacing a second element in a compound.* The three types of displacement reactions are: *a) an active metal displacing the metal from a less active metal's salt, b) an active metal displacing hydrogen from either HCl or H$_2$SO$_4$, and c) an active nonmetal displacing the nonmetal from a less active nonmetal's salt.*

Also, in this reaction Al is oxidized from the 0 to the +3 oxidation state and H is reduced from the +1 to the 0 oxidation state.

Displacement reactions involve the reaction of metals or nonmetals on the activity series with salts of less active metals or the nonoxidizing acids HCl and H$_2$SO$_4$.   HNO$_3$ is the most common oxidizing acid. If you are not familiar with the activity series in your text, be certain that you understand how to use it.

Metathesis Reaction Sample Exercise

5. ***What reaction types are represented by this reaction?***

$$Ba(OH)_2 \; + \; H_2SO_4 \rightarrow BaSO_4 \; + \; 2\,H_2O$$

The correct answer is a metathesis reaction that is both an acid-base and a precipitation reaction.

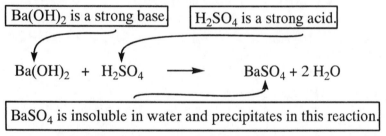

| Ba(OH)$_2$ is a strong base. | | H$_2$SO$_4$ is a strong acid. |

$$Ba(OH)_2 \; + \; H_2SO_4 \longrightarrow BaSO_4 + 2\,H_2O$$

BaSO$_4$ is insoluble in water and precipitates in this reaction.

INSIGHT: Metathesis reactions are characterized by *the reactants switching their anions.*

This is exhibited by using the symbols AB to represent one reactant and CD to represent the other reactant. The products are represented by AD and CB.

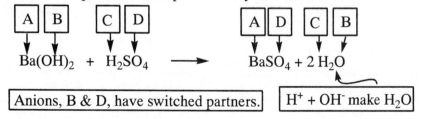

| A | B | | C | D | | | A | D | | C | B |

Ba(OH)$_2$ + H$_2$SO$_4$ ⟶ BaSO$_4$ + 2 H$_2$O

Anions, B & D, have switched partners. | H$^+$ + OH$^-$ make H$_2$O

When an acid reacts with a base, a salt will be formed, $BaSO_4$ in this case, and water if the base is a hydroxide. Water is formed by the combination of the H^+ with the OH^-.

INSIGHT: | Precipitation reactions are characterized by *the formation of a compound that is insoluble in water.*

You must understand and use the solubility rules from your textbook to recognize a precipitation reaction.

Predicting Products of Metathesis Reactions Sample Exercise

6. *What are the products of this chemical reaction?*

$$Sr(OH)_2 \ + \ Fe(NO_3)_3 \rightarrow ??? + ???$$

The correct answer is $Sr(NO_3)_2$ and $Fe(OH)_3$.

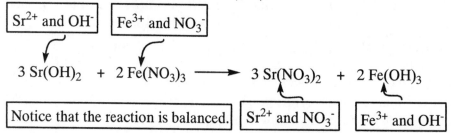

INSIGHT: | The anions have switched partners forming new chemical compounds. Basic rules of ionic compound formation must be obeyed, thus the total charge of the positive ions is equal to the total charge of the negative ions.

Total and Net Ionic Equations Sample Exercises

7. *Write the total ionic and net ionic equations for this reaction.*

$$Ba(OH)_{2(aq)} \ + \ 2\ HCl_{(aq)} \ \rightarrow \ BaCl_{2(aq)} \ + \ 2\ H_2O_{(\ell)}$$

The correct total ionic equation is:

$$Ba^{2+}_{(aq)} + 2\ OH^-_{(aq)} + 2\ H^+_{(aq)} + 2\ Cl^-_{(aq)} \rightarrow Ba^{2+}_{(aq)} + 2\ Cl^-_{(aq)} + 2\ H_2O_{(\ell)}$$

The correct net ionic equation is:

$$2\ OH^-_{(aq)} + 2\ H^+_{(aq)} \rightarrow 2\ H_2O_{(\ell)}$$

or

$$OH^-_{(aq)} + H^+_{(aq)} \rightarrow H_2O_{(\ell)}$$

Unless you are readily familiar with the solubility rules, these problems are exceedingly difficult.

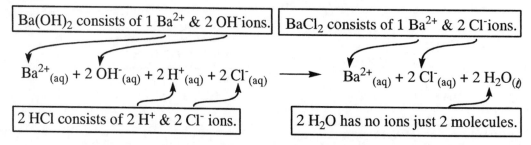

$$2\ OH^-_{(aq)} + 2\ H^+_{(aq)} \rightarrow 2\ H_2O_{(\ell)}$$
or
$$OH^-_{(aq)} + H^+_{(aq)} \rightarrow H_2O_{(\ell)}$$

8. **Write the total and net ionic equations for this reaction.**

$$NaOH_{(aq)} + CH_3COOH_{(aq)} \rightarrow NaCH_3COO_{(aq)} + H_2O_{(\ell)}$$

The correct total ionic equation is:

$$Na^+_{(aq)} + OH^-_{(aq)} + CH_3COOH_{(aq)} \rightarrow Na^+_{(aq)} + CH_3COO^-_{(aq)} + H_2O_{(\ell)}$$

The correct net ionic equation is:

$$OH^-_{(aq)} + CH_3COOH_{(aq)} \rightarrow CH_3COO^-_{(aq)} + H_2O_{(\ell)}$$

NaOH is a **strong** water soluble base that ionizes into Na^+ and OH^- ions in aqueous solutions.	NaCH_3COO is a **water soluble salt** that ionizes into Na^+ and CH_3COO^- ions in aqueous solutions.

$$Na^+_{(aq)} + OH^-_{(aq)} + CH_3COOH_{(aq)} \longrightarrow Na^+_{(aq)} + CH_3COO^-_{(aq)} + H_2O_{(\ell)}$$

CH_3COOH is a **weak** water soluble acid that ionizes so slightly in aqueous solutions that it is not separated into ions.	H_2O is a molecule and does not form ions.
The only spectator ion in this reaction is Na^+.	Removing the Na^+ ion from the total ionic equation leaves the net ionic equation.

$$OH^-_{(aq)} + CH_3COOH_{(aq)} \rightarrow CH_3COO^-_{(aq)} + H_2O_{(\ell)}$$

Notice that the net ionic equation tells us that a strong base, hydroxide ion, reacts with a weak acid, acetic acid, to form the acetate ion and water.

Module 7
Electronic Structure of Atoms

Introduction
This module will help you understand the basic rules of quantum numbers. The important things to understand are <u>how to determine the quantum numbers for an element, how to discern the correct atomic electronic structure of an element by looking at the periodic table, and how to write the entire set of quantum numbers for an element.</u>

Module 7 Key Equations & Concepts

1. $n = 1, 2, 3, 4, 5, 6, ...\infty$
   This equation defines the principal quantum number, n. The main energy level of an atom is described by its principal quantum number.

2. $\text{ł} = 0, 1, 2, 3, 4, (n-1)$ or $\text{ł} = s, p, d, f, g, ...(n-1)$
   This equation defines the orbital angular momentum quantum number, ł. The shape of the atomic orbitals, the region of space occupied by electrons, is described by the orbital angular momentum quantum number.

3. $m_{\text{ł}} = -\text{ł}, -\text{ł}+1, -\text{ł}+2, ..., 0, ... \text{ł}-2, \text{ł}-1, \text{ł}$
   This equation defines the magnetic quantum number, $m_{\text{ł}}$. The number of atomic orbitals that are possible for each value of ł is defined by the magnetic quantum number.

4. $m_s = +1/2$ and $-1/2$
   This equation defines the spin quantum number, m_s. The spin quantum number describes the relative magnetic orientation of the electrons in an atom. It also defines the maximum number of electrons that can occupy one orbital.

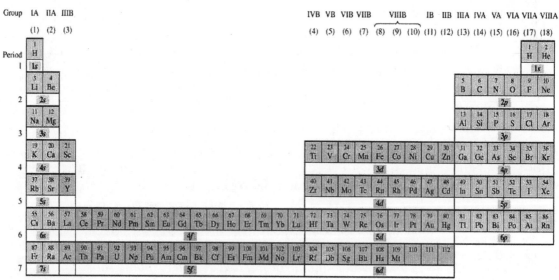

© 2004 Thomson/Brooks Cole

Principal Quantum Number Sample Exercises

1. What is the value of the principal quantum number, n, for the valence electrons in a Sr atom?

The correct answer is n = 5.

2. **What is the value of the n quantum number for the valence electrons in a Zr atom?**

The correct answer is n = 4.

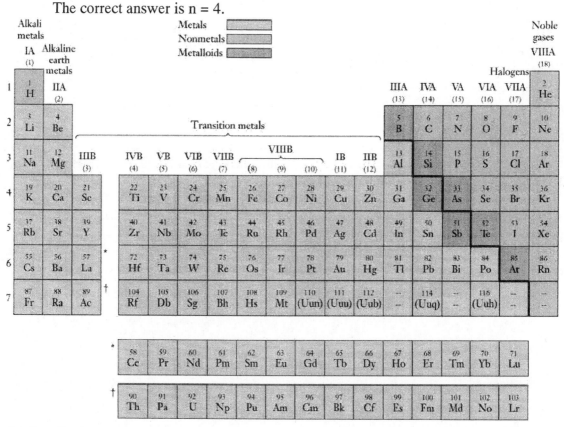

© 2004 Thomson/Brooks Cole

Orbital Angular Momentum Quantum Number Sample Exercises

3. **What is the value of the orbital angular momentum quantum number, ℓ, for the valence electrons in a Sr atom?**

The correct answer is ℓ = 0 or s electrons.

For Sr which has n = 5, ℓ can be 0, 1, 2, 3, 4 or s, p, d, f, g electrons.

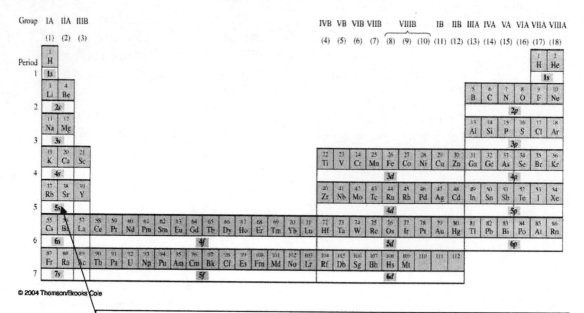

INSIGHT: Notice that the two electrons which make Sr, element 38, different from Ar, element 36, are in the s block of the periodic table. *All s electrons have an ℓ value of 0.*

4. What is the value of the orbital angular momentum quantum number, ℓ, for the electrons which make a Zr atom different from a Sr atom?

The correct answer is ℓ = 2 or d electrons.

For Sr which has n = 4, ℓ can be 0, 1, 2, 3 or s, p, d, f electrons.

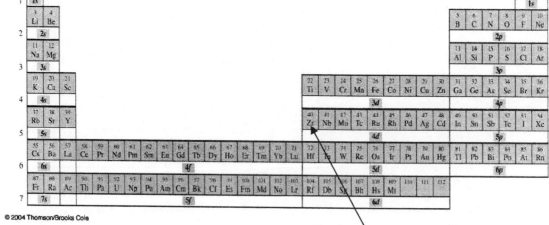

INSIGHT: The electrons which make a Zr atom, element 45, different from a Sr atom, element 38, are in the d block of the periodic table. *All d electrons have an ℓ value of 2.*

Magnetic Quantum Number Sample Exercises

5. *What is the value of the magnetic quantum number, m_ℓ, for the valence electrons in a Sr atom?*

    The correct answer is $m_\ell = 0$.

    For Sr which has an $\ell = 0$, the only possible value of m_ℓ is 0.

    INSIGHT: | For s electrons $\ell = 0$. The equation for the magnetic quantum number m_ℓ is $m_\ell = -\ell, -\ell+1, -\ell+2, ..., 0, ... \ell-2, \ell-1, \ell$. If $\ell = 0$, then $-\ell = -0$ and $+\ell = +0$, thus the only possible value of m_ℓ is 0.

6. *What is the value of the magnetic quantum number, m_ℓ, for the electrons which make a Zr atom different from a Sr atom?*

    The correct answer is $m_\ell = -2, -1, 0, +1, +2$.

    The electrons which make Zr different from Sr are d electrons having an $\ell = 2$.

    INSIGHT: | For $\ell = 2$, $-\ell = -2$, $-\ell + 1 = -1$, $-\ell + 2 = 0$, $\ell - 1 = 1$, $\ell = 2$. These five values of m_ℓ (-2, -1, 0, +1, +2 is five different numbers) indicate that there are five different d orbitals. *You must think of quantum numbers as labels rather than numbers.*

Spin Quantum Number Sample Exercise

7. *What is the value of the spin quantum number, m_s, for the valence electrons in a Sr atom?*

    The correct answer is $m_s = +1/2$ and $-1/2$.

    One of the 5s electrons in Sr will have $m_s = +1/2$ the other will have $m_s = -1/2$.

    INSIGHT: | m_s can only have two possible values, +1/2 and -1/2.

Electronic Structure from the Periodic Chart Sample Exercises

8. *What is the correct electronic structure of the Sr atom? Write the structure in both orbital notation and simplified (or spdf) notation.*

    The correct answer is [Kr] $\underline{\uparrow\downarrow}$ or [Kr] $5s^2$.
    $5s$

    INSIGHT: | The noble gas core configuration is the determined by starting at the element and decreasing the atomic number until reaching a noble gas.

27

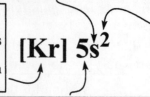

s is the orbital angular momentum quantum number because Sr's valence electrons are in the s block of the periodic table.

This symbol indicates that 36 of the 38 electrons in Sr are in the same orbitals as in the noble gas Kr.

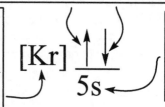

$[Kr]\ 5s^2$

2 indicates that both of the distinguishing electrons in Sr are s electrons.

5 is the principal quantum number because Sr is on the 5th row of the periodic table.

The ↑ indicates that the m_s for one of the valence electrons is +1/2. The ↓ symbolizes that m_s = -1/2 for the second valence electron.

As above, this symbol indicates that 36 of the 38 electrons in Sr are in the same orbitals as in the noble gas Kr.

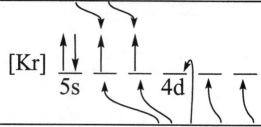

$[Kr]$

5s

5s is the symbol for the **n** and l quantum numbers for Sr.

9. *What is the correct electronic structure of the Zr atom? Write the structure in both orbital notation and simplified (or spdf) notation.*

The correct answer is [Kr] ↑↓ ↑ ↑ __ __ __ or $[Kr]\ 5s^2\ 4d^2$.
    5s   4d

This symbol represents the two d electrons that differentiate Zr from Sr.

$[Kr]\ 5s^2\ 4d^2$

4d is the symbol for the **n** and l quantum numbers for the d electrons in Zr.

These two arrows are not paired because of Hund's rule.

[Kr] ↑↓ ↑ ↑ __ __ __
    5s   4d

The five spaces above the 4d symbol represent the five d orbitals that are possible in the 4d energy level.

28

Writing Quantum Numbers Sample Exercises

10. *Write the correct set of quantum numbers for the valence electrons in Sr.*

The correct answer is:

n	l	m_l	m_s	
5	0	0	+1/2	1st electron quantum numbers
5	0	0	-1/2	2nd electron quantum numbers

INSIGHT: If both of the m_s numbers were reversed, the answer would also be correct.

11. *Write the correct set of quantum numbers for the valence electrons in Zr.*

The correct answer is:

n	l	m_l	m_s	
5	0	0	+1/2	1st electron quantum numbers
5	0	0	-1/2	2nd electron quantum numbers
4	2	-2	+1/2	3rd electron quantum numbers
4	2	-1	+1/2	4th electron quantum numbers

INSIGHT: Strictly speaking, m_l could be any two of the five possible values and be correct.

INSIGHT: Both of the m_s values for the 4d's could also be -1/2 and be correct. But having one value +1/2 and the other -1/2 is incorrect because it does not obey Hund's rule.

 YIELD There are several important rules that you need to know to understand electron configurations. These include Hund's rule, the Pauli Exclusion Principle, and the Aufbau Principle. Be certain that you know and understand these rules. The most important thing to learn from this module is how to get the correct electronic configuration of an element using the periodic table.

Module 8
Chemical Periodicity

Introduction

There are many properties of elements that are based upon their electronic structures. Using some very simple rules, we can predict the variations of some properties based upon the element's position on the periodic chart. This module will help you learn the periodic properties of some important chemical properties. The important points to learn from this module are <u>the periodic properties associated with electronegativity, ionization energy, electron affinity, atomic radii, and ionic radii</u>. You will need to have a periodic chart with you as you work on this module because you must learn to associate these properties on the periodic chart.

Module 8 Key Equations & Concepts

1. *Electronegativity* **is the relative measure of an element's ability to attract electrons to itself.**

    Electronegativity helps us determine the likelihood of ionic or covalent bond formation and the polarity of molecules.

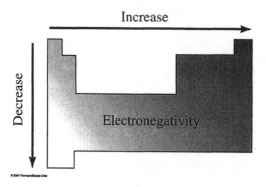

2. *Ionization energy* **is the amount of energy required to remove an electron from an atom or ion.**

    Ionization energy is an important indicator of an element's likelihood of forming positive ions. Elements with several electrons can have a 1st ionization energy, 2nd ionization energy, and so forth until all of that element's electrons have been removed.

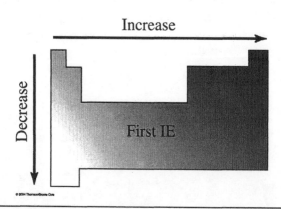

30

3. *Electron affinity* **is the amount of energy absorbed when an electron is added to an isolated gaseous atom.**

Electron affinity will help us understand which elements are most likely to form negative ions. Electron affinity has the most irregular periodic trends of the properties discussed in this module.

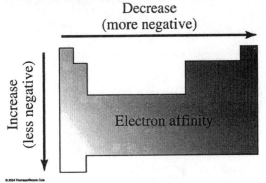

4. *Atomic radii* **are the measured distances from the center of the atom to its outer electrons.**

Atomic radii will help us predict the solid state structure of the elements.

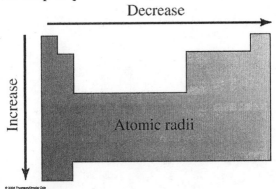

5. *Ionic radii* **are the measured distances from the center of an ion to its outer electrons.**

There are ionic radii trends for both positive and negative ions. These trends will help us determine the strength of ionic bonds. Note that cations are always smaller than their parent atoms and anions are always bigger than their parent atoms.

IA	IIA		IIIA	IVA	VA	VIA	VIIA	VIIIA
Li⁺ 0.90	Be²⁺ 0.59	Ionic radii			N³⁻ 1.71	O²⁻ 1.26	F⁻ 1.19	
Na⁺ 1.16	Mg²⁺ 0.85		Al³⁺ 0.68			S²⁻ 1.70	Cl⁻ 1.67	
K⁺ 1.52	Ca²⁺ 1.14		Ga³⁺ 0.76			Se²⁻ 1.84	Br⁻ 1.82	
Rb⁺ 1.66	Sr²⁺ 1.32		In³⁺ 0.94			Te²⁻ 2.07	I⁻ 2.06	
Cs⁺ 1.81	Ba²⁺ 1.49		Tl³⁺ 1.03				2 Å	

31

Electronegativity Sample Exercise

1. *Arrange these elements from smallest to largest based on their electronegativity.*
   O, Ca, Si, Cs

The correct answer is Cs < Ca < Si < O.

> The most electronegative elements are in the upper right corner of the periodic chart.

Electronegativities of the Elements

	Metals	▓▓▓
	Nonmetals	☐
	Metalloids	▓▓▓

	IA																		VIIIA
1	1 H 2.1	IIA											IIIA	IVA	VA	VIA	VIIA	2 He
2	3 Li 1.0	4 Be 1.5											5 B 2.0	6 C 2.5	7 N 3.0	8 O 3.5	9 F 4.0	10 Ne
3	11 Na 1.0	12 Mg 1.2	IIIB	IVB	VB	VIB	VIIB	VIIIB			IB	IIB	13 Al 1.5	14 Si 1.8	15 P 2.1	16 S 2.5	17 Cl 3.0	18 Ar
4	19 K 0.9	20 Ca 1.0	21 Sc 1.3	22 Ti 1.4	23 V 1.5	24 Cr 1.6	25 Mn 1.6	26 Fe 1.7 27 Co 1.7 28 Ni 1.8			29 Cu 1.8	30 Zn 1.6	31 Ga 1.7	32 Ge 1.9	33 As 2.1	34 Se 2.4	35 Br 2.8	36 Kr
5	37 Rb 0.9	38 Sr 1.0	39 Y 1.2	40 Zr 1.3	41 Nb 1.5	42 Mo 1.6	43 Tc 1.7	44 Ru 1.8 45 Rh 1.8 46 Pd 1.8			47 Ag 1.6	48 Cd 1.6	49 In 1.6	50 Sn 1.8	51 Sb 1.9	52 Te 2.1	53 I 2.5	54 Xe
6	55 Cs 0.8	56 Ba 1.0	57 La 1.1 *	72 Hf 1.3	73 Ta 1.4	74 W 1.5	75 Re 1.7	76 Os 1.9 77 Ir 1.9 78 Pt 1.8			79 Au 1.9	80 Hg 1.7	81 Tl 1.6	82 Pb 1.7	83 Bi 1.8	84 Po 1.9	85 At 2.1	86 Rn
7	87 Fr 0.8	88 Ra 1.0	89 Ac 1.1 †															

*
58 Ce 1.1	59 Pr 1.1	60 Nd 1.1	61 Pm 1.1	62 Sm 1.1	63 Eu 1.1	64 Gd 1.1	65 Tb 1.1	66 Dy 1.1	67 Ho 1.1	68 Er 1.1	69 Tm 1.1	70 Yb 1.0	71 Lu 1.2

†
90 Th 1.2	91 Pa 1.3	92 U 1.5	93 Np 1.3	94 Pu 1.3	95 Am 1.3	96 Cm 1.3	97 Bk 1.3	98 Cf 1.3	99 Es 1.3	100 Fm 1.3	101 Md 1.3	102 No 1.3	103 Lr 1.5

© 2004 Thomson/Brooks Cole

> The least electronegative elements are in the lower left corner of the periodic chart.

> Electronegativity steadily increases moving from the lower left to the upper right corners of the periodic chart.

Ionization Energy Sample Exercise

2. *Arrange these elements from smallest to largest based on their ionization energies.*

F, N, C, O

The correct answer is C < O < N < F.

First ionization energies increase steadily from the alkali metals to the noble gases.

Elements with filled s orbitals.

Elements with half-filled p orbitals.

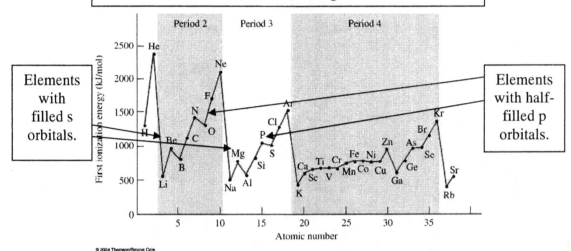

There are significant variations from the steady increase at the IIA metals and the VA nonmetals due to filled s orbitals or half-filled p orbitals.

Electron Affinity Sample Exercise

3. *Arrange these elements from smallest to largest based on their electron affinities.*

F, N, C, O

The correct answer is F < O < C < N.

Notice that elements with filled or half-filled electron sub shells have the largest electron affinities.

N, having half-filled p orbitals, has a slightly positive electron affinity indicating that it less easily forms anions.

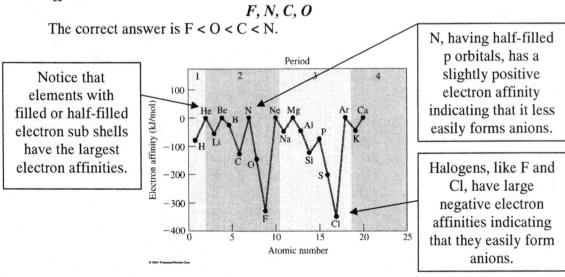

Halogens, like F and Cl, have large negative electron affinities indicating that they easily form anions.

Atomic Radii Sample Exercise

4. *Arrange these elements from smallest to largest based on their atomic radii.*

$$F, Ga, S, Rb$$

The correct answer is F < S < Ga < Rb.

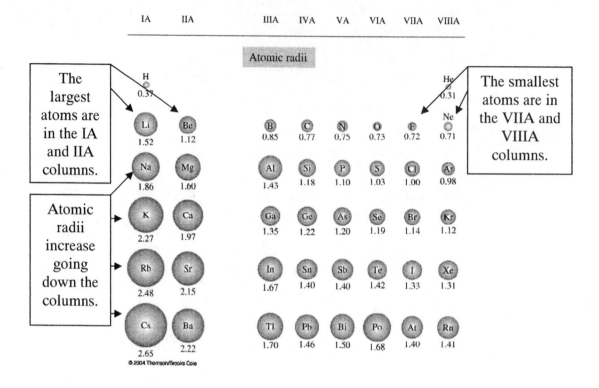

Ionic Radii Sample Exercise

5. *Arrange these ions from smallest to largest based on their ionic radii.*

$$S^{2-}, Cl^-, Mg^{2+}, Al^{3+}$$

The correct answer is $Al^{3+} < Mg^{2+} < Cl^- <. S^{2-}$.

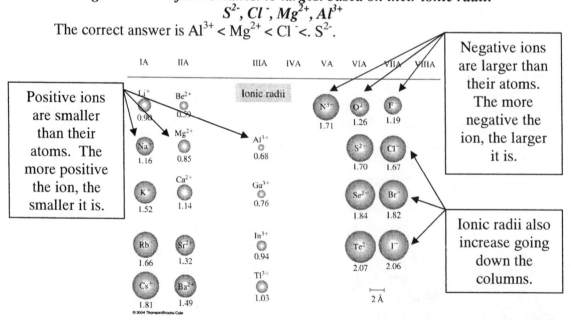

Module 9
Chemical Bonding

Introduction

This module is concerned with how chemical bonds are formed. There are two basic types of chemical bonds, ionic and covalent. This module's important points are how to determine if a compound is ionic or covalent, drawing Lewis dot structures of atoms, writing formulas of the simple ionic compounds, determining relative ionic bond strength, drawing Lewis dot structures of ionic and covalent compounds, and recognizing if a covalent bond is polar or nonpolar. A periodic table will help you understand many of the electron structures used in this chapter.

Module 9 Key Equations & Concepts

1.
$$\text{Force of attraction between 2 ions} \propto \frac{q^+ \times q^-}{d^2}$$

$$\text{Force of attraction between 2 ions} \propto \frac{(\text{Charge on cation})(\text{Charge on anion})}{(\text{Distance between the ions})^2}$$

Coulomb's Law describes the strength of attraction between two ions of opposite charge. This will help us determine the strength of ionic bonds.

Determining if a Compound is Ionic or Covalent Sample Exercise

1. *Indicate which of the following compounds are ionic in nature and which are covalent in nature.*

$$CH_4, KBr, Ca_3N_2, Cl_2O_7, H_2SO_4, InCl_3$$

The correct answer is KBr, Ca_3N_2 and $InCl_3$ are ionic while CH_4, Cl_2O_7, and H_2SO_4 are covalent.

INSIGHT: | Ionic compounds are formed by the reaction of metallic elements with nonmetallic elements or the reaction of the ammonium ion, NH_4^+, with nonmetals.

| K is a metallic element. | Ca is a metallic element. | In is a metallic element. |

KBr   Ca₃N₂   InCl₃

| Br is a nonmetallic element. | N is a nonmetallic element. | Cl is a nonmetallic element. |

INSIGHT: | Covalent compounds are formed by the reaction of nonmetals with nonmetals.

| C is a nonmetal. | Cl is a nonmetal. | H and S are nonmetals. |

CH₄   Cl₂O₇   H₂SO₄

| H is a nonmetal. | O is a nonmetal. | O is a nonmetal. |

Lewis Dot Structures of Atoms Sample Exercises

2. *Draw the correct Lewis dot structure of these elements.*
   Mg, P, S, Ar

The correct answer is:

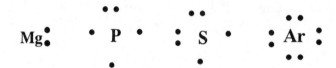

Lewis dot structures reflect the electronic structures of the elements, including how the electrons are paired. Notice how the orbital diagrams match the Lewis dot structures of each element.

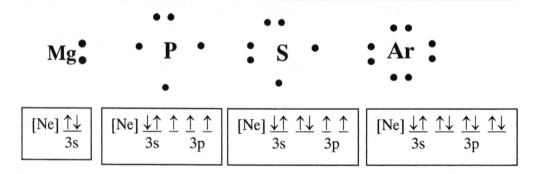

[Ne] $\uparrow\downarrow$	[Ne] $\downarrow\uparrow$ $\uparrow$ $\uparrow$ $\uparrow$	[Ne] $\downarrow\uparrow$ $\uparrow\downarrow$ $\uparrow$ $\uparrow$	[Ne] $\downarrow\uparrow$ $\uparrow\downarrow$ $\uparrow\downarrow$ $\uparrow\downarrow$
3s	3s 3p	3s 3p	3s 3p

Simple Ionic Compounds Sample Exercise

3. *Write the correct formulas of the ionic compounds formed when a) Mg atoms react with Cl atoms, b) Mg atoms react with S atoms, c) Mg atoms react with P atoms.*

The correct answers are: $MgCl_2$, MgS, and Mg_3P_2.

Mg, like all of the IIA metals, has two electrons in its valence shell and commonly forms +2 ions, Mg^{2+}.		
$MgCl_2$	MgS	Mg_3P_2
Cl, and all of the VIIA nonmetals, have seven electrons in their valence shell and commonly form 1- ions, Cl^-. Two Cl^- ions have the same size charge as one Mg^{2+} ion.	S, and all of the VIA nonmetals, have six electrons in their valence shell and commonly form 2- ions, S^{2-}. One S^{2-} ion has the same size charge as one Mg^{2+} ion.	P, and all of the VA nonmetals, have five electrons in their valence shell and commonly form 3- ions, P^{3-}. Two P^{3-} ions have the same size charge as three Mg^{2+} ions.

4. *Arrange these three ionic compounds, from smallest to largest, based on the strength of the ionic bonds in each compound.*
      MgSe, MgO, MgS
   The correct answer is MgSe < MgS < MgO.

Coulomb's Law tells us that the force of attraction between ions $= \dfrac{q^+ \times q^-}{d^2}$. The strongest ionic bond will have the largest charge with the smallest ionic radii. Module 8 discusses the periodicity of ionic radii.

MgSe	<	**MgS**	<	**MgO**
Mg^{2+} and Se^{1-} are the largest pair of ions.		Mg^{2+} and S^{2-} are the medium sized pair of ions.		Mg^{2+} and O^{2-} are the smallest pair of ions.

Drawing Lewis Dot Structures of Ionic Compounds Sample Exercise
5. *Draw the Lewis dot structures for each of these compounds.*
      AlP, NaCl, MgCl₂
   The correct answers are:

$$Al^{3+} \left[: \ddot{P} : \right]^{3-} \qquad Na^+ \left[: \ddot{Cl} : \right]^- \qquad Mg^{2+} \; 2\left[: \ddot{Cl} : \right]^-$$

Al loses all three of its valence electrons, thus has no dots, and forms a 3+ ion.

$$Al^{3+} \left[: \ddot{P} : \right]^{3-}$$

P gains three electrons from Al, thus has 8 dots (5 valence electron plus 3 from Al), and forms a 3- ion. The []'s indicate that the 3- charge is associated with the P ion.

Na loses its one valence electron, having no dots, and forms a 1+ ion.

$$Na^+ \left[: \ddot{Cl} : \right]^-$$

Cl gains one electron from Na, thus has 8 dots (7 valence electron plus 1 from Na), and forms a 1- ion.

Mg loses both valence electrons, having no dots, and forms a 2+ ion.

$$Mg^{2+} \; 2\left[: \ddot{Cl} : \right]^-$$

Each Cl atom gains one electron from the Mg. The 2 in front of the []'s indicates that two Cl⁻ ions are needed to have the same size charge as the Mg^{2+}.

Simple Covalent Compounds Sample Exercise

6. Draw the correct Lewis dot structures for each of these compounds.
SiH_4, PCl_3, SF_6

The correct answers are:

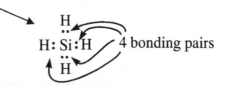

In most cases every element in a compound will obey the octet rule. Thus Si has a share of 8 electrons and each H has a share of 2 electrons. This compound has only *bonding pairs* of electrons.

In this compound P has a share of 8 electrons and each Cl has a share of 8 electrons. This compound has 3 *bonding pairs* and 1 *lone pair* of electrons.

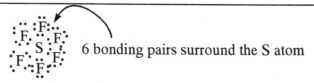

This compound does not obey the octet rule.
S has a share of 12 electrons while each F has a share of 8 electrons. This compound has 6 *bonding pairs* of electrons. Look in your textbook for the rules on which compounds do not obey the octet rule.

6 bonding pairs surround the S atom

INSIGHT: When drawing Lewis dot structures, if the compound obeys the octet rule, the central atom will have a share of 8 electrons. The possible combinations of 8 electrons for compounds that **obey the octet rule** are:

Bonding Pairs	Lone Pairs
4	0
3	1
2	2
1	3

38

If the compound **does not obey the octet rule**, the central atom can have 2, 3, 5 or 6 pairs of electrons around the central atom. All of the possible combinations of 2, 3, 5, or 6 pairs of electrons are:

Total Pairs of Electrons	Bonding Pairs	Lone Pairs
2	2	0
3	3	0
5	5	0
5	4	1
5	3	2
5	2	3
6	6	0
6	5	1
6	4	2

INSIGHT: The noncentral atoms will obey the octet rule having either 1 bonding pair, as for H atoms, or a share of 8 electrons as is the case for Cl and F in the examples above.

Polar or Nonpolar Covalent Bonds in Compounds Sample Exercise

7. *Which of these compounds contains polar covalent bonds?*

$$F_2, CH_4, H_2O$$

The correct answer is CH_4 and H_2O contain polar covalent bonds and F_2 does not.

The periodic trends regarding electronegativity are discussed in Module 8. You will need to know those trends for problems of this nature.

Polar bonds occur when the two atoms involved in the bond have a difference in electronegativity. In F_2 the two atoms are both F. They have the same electronegativity thus there is not a polar bond. In CH_4 and H_2O, the H to the central atom (C or O) bond involves atoms with different electronegativities. Thus there are polar covalent bonds in CH_4 and H_2O.

Module 10
Molecular Shapes

Introduction

Molecular shape refers to the geometrical arrangement of atoms around the central atom in a molecule or polyatomic ion. This module will help you understand and predict the stereochemistry of many molecules. The most important ideas to take from this module are how to predict and name the geometrical shapes of molecules. Molecular shapes are important in the chemical reactivity of numerous compounds.

Module 10 Key Equations & Concepts

1. **Molecules with <u>2 regions of high electron density</u> have** *linear* **electronic and molecular geometries** at the central atom with 180° bond angles.
   Examples include BeF_2, BeH_2, $BeCl_2$, and $BeBr_2$.

2. **Molecules with <u>3 regions of high electron density</u> have** *trigonal planar* **electronic and molecular geometries** at the central atom with 120° bond angles.
   Examples include BH_3, BF_3, $AlCl_3$, and GaI_3.

3. **Molecules with <u>4 regions of high electron density</u> have** t*etrahedral* **electronic geometry and can have either tetrahedral, trigonal pyramidal, angular, or linear molecular geometries** at the central atom. Bond angles vary with the molecular geometries.
   Examples include CH_4, SiH_4, PF_3, and H_2O.

4. **Molecules with <u>5 regions of high electron density</u> have** *trigonal bipyramidal* **electronic geometry and can have either trigonal bipyramidal, see-saw, T-shaped, or linear molecular geometries** at the central atom. Bond angles vary with the molecular geometries.
   Examples include PF_5, SF_4, ClF_3, and XeF_2.

5. **Molecules with <u>6 regions of high electron density</u> have** *octahedral* **electronic geometry and can have either octahedral, square-based pyramidal, or square-planar molecular geometrie**s at the central atom. Bond angles vary with the molecular geometries.
   Examples include SF_6, IF_5, and XeF_4.

Electronic Geometry versus Molecular Geometry

All of the molecules described in this module have two geometries that you must be familiar with, their electronic and molecular geometries.

1) **Electronic geometry** considers all of the regions of high electron density including bonding pairs, lone pairs, and double or triple bonds.

2) **Molecular geometry** only considers those electrons and atoms that are involved in bonding pairs or in double and triple bonds.

The molecular geometry will be different from the electronic geometry only in molecules that have lone pairs of electrons.

Number of Regions of High Electron Density	Electronic Geometry*		
	Description; Angles†	Line Drawing‡	Ball-and-stick model
2	linear; 180°		
3	trigonal planar; 120°		
4	tetrahedral; 109.5°		
5	trigonal bipyramidal; 90°, 120°, 180°		
6	octahedral; 90°, 180°		

Linear Molecules Sample Exercise

1. What are the correct molecular geometries of these molecules?

$$BeI_2, BeHF$$

The correct answer is linear for both molecules.

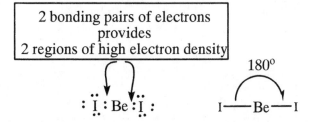

2 bonding pairs of electrons provides
2 regions of high electron density

180°

$: \ddot{I} : Be : \ddot{I} :$ $I \text{---} Be \text{---} I$

Notice that the linear shape is determined by the electrons around the central Be atom, not the lone pairs on the I atoms.

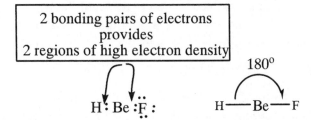

2 bonding pairs of electrons provides
2 regions of high electron density

180°

$H : Be : \ddot{F} :$ $H \text{---} Be \text{---} F$

The linear shape is not affected by the two different atoms bonded to Be. The shape is determined by the 2 regions of high electron density.

INSIGHT: Covalent Compounds of Be do not obey the octet rule. If Be is the central atom in a molecule there will be 2 regions of high electron density and the electronic and molecular geometries will be linear.

Trigonal Planar Molecules Sample Exercise

    2. What are the correct molecular geometries of these molecules?

$$BH_3, AlHFBr$$

The correct answer is trigonal planar for both molecules.

┌─────────────────────────────────────┐
│ 3 bonding pairs of electrons   │
│ give   │
│ 3 regions of high electron density   │
└─────────────────────────────────────┘

$120°$ H $120°$

B

H   H

$120°$

┌─────────────────────────────────────┐
│ 3 bonding pairs of electrons   │
│ give   │
│ 3 regions of high electron density   │
└─────────────────────────────────────┘

$120°$ H $120°$

Al

Br   F

$120°$

┌──┐
│ Again, having 3 different atoms does not affect the molecule's shape. Shapes are │
│ determined by the regions of high electron density.   │
└──┘

INSIGHT: ┌──┐
│ Covalent compounds of the IIIA group (B, Al, Ga, & In) do not obey the │
│ octet rule. If a IIIA element is the central atom the molecule will have 3 │
│ regions of high electron density around the central atom and the │
│ electronic and molecular geometries will be trigonal planar. These │
│ molecules have 3 bonding pairs of electrons. │
└──┘

Tetrahedral and Variations of Tetrahedral Sample Exercise

    3. What are the correct molecular geometries of these molecules?

$$SiH_4, PF_3, H_2O$$

The correct answer is tetrahedral for SiH_4, trigonal pyramidal for PH_3, and bent or angular for H_2O.

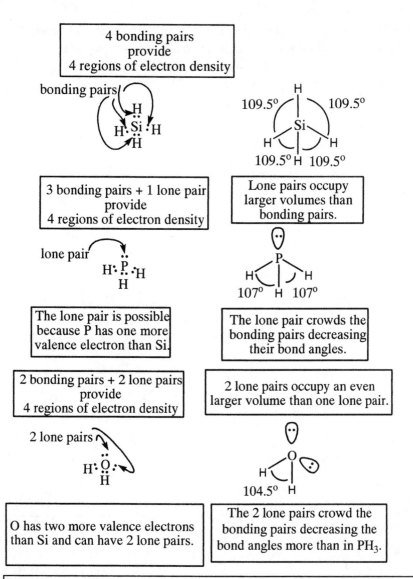

4 bonding pairs provide 4 regions of electron density

bonding pairs

Lone pairs occupy larger volumes than bonding pairs.

3 bonding pairs + 1 lone pair provide 4 regions of electron density

lone pair

$107°$ H $107°$

The lone pair is possible because P has one more valence electron than Si.

The lone pair crowds the bonding pairs decreasing their bond angles.

2 bonding pairs + 2 lone pairs provide 4 regions of electron density

2 lone pairs occupy an even larger volume than one lone pair.

2 lone pairs

$104.5°$ H

O has two more valence electrons than Si and can have 2 lone pairs.

The 2 lone pairs crowd the bonding pairs decreasing the bond angles more than in PH_3.

INSIGHT:

All of the molecules in this category will obey the octet rule because they have 4 regions of high electron density which is equivalent to 8 electrons.

1. If a IVA element (C, Si, or Ge) is the central atom, the electronic and molecular geometries will be tetrahedral. These molecules contain 4 bonding pairs of electrons.

2. If a VA element (N, P, or As) is the central atom, the electronic geometry will be tetrahedral and the molecular geometry will be trigonal pyramidal. These molecules contain 3 bonding pairs and 1 lone pair of electrons.

3. If a VIA element (O, S, Se) is the central atom, the electronic geometry will be tetrahedral and the molecular geometry will be bent, angular, or V-shaped. These molecules contain 2 bonding pairs and 2 lone pairs of electrons.

4. If a VIIA element (F, Cl, Br, or I) is the central atom, the electronic geometry will be tetrahedral and the molecular geometry will be linear. These molecules contain 1 bonding pair and 3 lone pairs of electrons.

Trigonal Bipyramidal and Variations of Trigonal Bipyramidal Sample Exercise
   4. What are the correct molecular geometries of these molecules?
                    PF_5, SF_4, ClF_3, XeF_2
   The correct answer is trigonal bipyramidal for PF_5, see-saw shaped for SF_4, T-shaped for ClF_3, and linear for XeF_2.

5 bonding pairs on P atom provide 5 regions of electron density

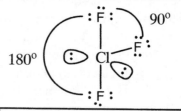

4 bonding pairs + 1 lone pair provide 5 regions of electron density	The increased volume of the lone pair on S changes the bond angles between the F atoms.

lone pair

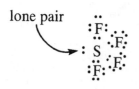

S has one more valence electron than P which makes the lone pair.	The see-saw shape is a simple modification of trigonal bipyramid due to the lone pair.

3 bonding pairs + 2 lone pairs provide 5 regions of electron density	Notice that both lone pairs occupy equatorial positions.

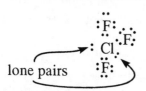

lone pairs

Cl has two more valence electrons than P which makes the two lone pairs.	The T- shape is another modification of trigonal bipyramid due to two lone pairs.

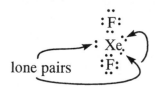

2 bonding pairs + 3 lone pairs provide
5 regions of electron density

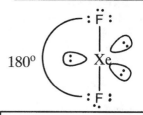

Notice that all three lone pairs occupy equatorial positions.

lone pairs

180°

Xe has three more valence electrons than P which makes the three lone pairs.

This linearshape is another modification of trigonal bipyramid due to three lone pairs.

<> YIELD <>

How do you recognize that there are 5 regions of high electron density or 10 electrons around the central atom?
Add the number of electrons in the valence shell of the central atom to the one electron that each of the noncentral atoms contributes to the central atom's orbitals. For example, in ClF_3 the central atom is Cl having 7 valence electrons. There are 3 F atoms each contributing 1 electron (the unpaired electron) to the Cl atom's orbitals. Thus the total is 7 for the Cl plus 3 x 1 for the F atoms = 10.

INSIGHT:

All of the molecules in this category do not obey the octet rule. They will have 5 regions of high electron density or 10 total electrons around the central atom.
1. If a VA element (P or As) is the central atom, the electronic and molecular geometries will be trigonal bipyramidal. These molecules contain 5 bonding pairs of electrons.
2. If a VIA element (S or Se) is the central atom, the electronic geometry will be trigonal bipyramidal and the molecular geometry will be seesaw. These molecules contain 4 bonding pairs and 1 lone pair of electrons.
3. If a VIIA element (Cl, Br, or I) is the central atom, the electronic geometry will be trigonal bipyramidal and the molecular geometry will be T-shaped. These molecules contain 3 bonding pairs and 2 lone pairs of electrons.
4. If an VIIIA element (Xe or Kr) is the central atom, the electronic geometry will be trigonal bipyramidal and the molecular geometry will be linear. These molecules contain 2 bonding pairs and 3 lone pairs of electrons.

Octahedral and Variations of Octahedral Sample Exercise
   5. What are the correct molecular geometries of these molecules?
   SF_6, IF_5, XeF_4

The correct answer is octahedral for SF_6, square-based pyramid for IF_5, and square planar for XeF_4.

6 bonding pairs on S provides 6 regions of electron density

90°   180°

5 bonding pairs and 1 lone pair provides 6 regions of electron density	The square-based pyramid shape is the octahedral shape with one lone pair of electrons.

lone pair

I has one more valence electron than S making the lone pair possible.

90°

4 bonding pairs and 2 lone pairs provides 6 regions of electron density	The square planar shape is the octahedral shape with two lone pairs of electrons.

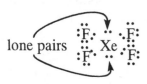

lone pairs

90°

Xe has two more valence electrons than S making the two lone pairs possible.

How do you recognize that there are 6 regions of high electron density or 12 electrons around the central atom? Just as for the previous sample exercise, add the valence electrons on the central atom to the 1 electron that each substituent provides to the central atom. For IF_5 that would be 7 electrons for the I plus 5 x 1 for the F atoms = 12 electrons total.

YIELD

INSIGHT:

All of the molecules in this category do not obey the octet rule. They will have 6 regions of high electron density or 12 total electrons around the central atom.

1. If a VIA element (S or Se) is the central atom, the electronic and the molecular geometries will be octahedral. These molecules contain 6 bonding pairs of electrons.

2. If a VIIA element (Cl, Br, or I) is the central atom, the electronic geometry will be octahedral and the molecular geometry will be square pyramidal. These molecules contain 5 bonding pairs and 1 lone pair of electrons.

3. If an VIIIA element (Xe or Kr) is the central atom, the electronic geometry will be octahedral and the molecular geometry will be square planar. These molecules contain 4 bonding pairs and 2 lone pairs of electrons.

Module 11
Hybridization and Polarity of Molecules

Introduction

Valence Bond theory is a different description of the shapes of molecules. It involves the hybridization (mixing) of atomic orbitals. The names of the orbitals come from the orbitals that have been mixed to make the shape. This module will help you understand and predict the hybridization of the atoms in several molecules. Polarity refers to whether the electron density of a molecule is symmetrically or asymmetrically arranged about the molecule. The most important ideas to take from this module are how to predict the hybridization of atoms using Valence Bond theory, how to understand the hybridization and geometry of double and triple bonds, and how to determine the polarity of a molecule.

Module 11 Key Equations & Concepts

1. **sp hybridized atoms**

   Atoms having two regions of electron density and a linear electronic geometry can be described as having orbitals that are made from one s and one p orbital.

2. **sp^2 hybridized atoms**

   Atoms having three regions of electron density and a trigonal planar electronic geometry can be described as having orbitals that are made from one s and two p orbitals.

3. **sp^3 hybridized atoms**

   Atoms having four regions of electron density and a tetrahedral electronic geometry can be described as having orbitals that are made from one s and three p orbitals.

4. **sp^3d hybridized atoms**

   Atoms having five regions of electron density and a trigonal bipyramidal electronic geometry can be described as having orbitals that are made from one s, three p, and one d orbitals.

5. **sp^3d^2 hybridized atoms**

   Atoms having six regions of electron density and an octahedral electronic geometry can be described as having orbitals that are made from one s, three p, and two d orbitals.

   The number of regions of electron density describes both the electronic geometry and the hybridization.

Regions of Electron Density	Electronic Geometry	Hybridization
2	Linear	sp
3	Trigonal planar	sp^2
4	Tetrahedral	sp^3
5	Trigonal bipyramidal	sp^3d
6	Octahedral	sp^3d^2

Hybridization Sample Exercises
   1. What is the hybridization of the underlined atom in each of these molecules?
$$\underline{Be}I_2, \underline{B}H_3, \underline{Si}H_4, \underline{P}F_5, \underline{S}F_6$$

The correct answer is sp hybridized for Be, sp^2 hybridized for B, sp^3 for Si, sp^3d for P, sp^3d^2 for S.

INSIGHT:

The names of the hybrid orbitals are derived from the orbitals used to make the hybrid.
1 s orbital mixed with 2 p orbitals

$$sp^2$$

<u>1</u> s orbital mixed with <u>2</u> p orbitals
yields <u>3</u> sp^2 orbitals

$$\underline{Be}I_2$$

2 regions of electron density   Be atom is sp hybridized

I——Be——I   ⌒Be⌒

1 s orbital + 1 p orbital = 2 sp hybrid orbitals

INSIGHT: Any central atom that has **2 regions of high electron density and a linear electronic geometry** can be described as an **sp hybrid**. See Sample Exercises 2 and 3 below for examples of different central atoms.

$$\underline{B}H_3$$

3 regions of electron density   B atom is sp^2 hybridized

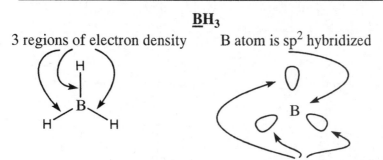

1 s orbital + 2 p orbitals = 3 sp^2 hybrid orbitals

INSIGHT: Any central atom that has **3 regions of high electron density and a trigonal planar electronic geometry** can be described as an **sp^2 hybrid**. See Sample Exercises 2 and 3 below for examples of different central atoms.

<u>SiH</u>$_4$

4 regions of electron density   Si atom is sp^3 hybridized

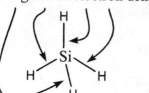 

1 s orbital + 3 p orbitals = 4 sp^3 hybrid orbitals

INSIGHT: Any central atom that has **4 regions of high electron density and a tetrahedral electronic geometry** can be described as an **sp^3 hybrid**. See Sample Exercises 2 and 3 below for examples of different central atoms.

<u>PF</u>$_5$

5 regions of electron density   P atom is sp^{3}d hybridized

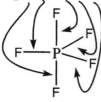 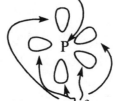

1 s orbital + 3 p orbitals + 1 d orbital = 5 sp^{3}d hybrid orbitals

INSIGHT: Atoms that have **5 regions of high electron density and a trigonal bipyramidal electronic geometry** can be described as an **sp^{3}d hybrid**. To form this hybrid an atom must have available empty d orbitals. Thus only central atoms on the third to the sixth row of the periodic chart can form sp^{3}d hybrid orbitals. Common examples of molecules containing central atoms that are sp^{3}d hybridized are SF$_4$, ClF$_3$, and XeF$_2$.

<u>SF</u>$_6$

6 regions of electron density   S atom is sp^{3}d^2 hybridized

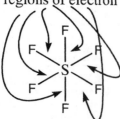 

1 s orbital + 3 p orbitals + 2 d orbitals = 6 sp^{3}d^2 hybrid orbitals

Hybridization of Double and Triple Bonds Sample Exercises

2. *What is the hybridization of the underlined atoms in these molecules?*
   $\underline{C}_2H_4$, $\underline{C}_2H_2$, $H_2C\underline{O}$

The correct answer is sp^2 hybridized for the C's in C_2H_4, sp hybridized for the C's in C_2H_2, and sp^2 hybridized for the O in H_2CO.

The key to answering questions involving geometry and hybridization in compounds containing double and triple bonds is counting the regions of high electron density surrounding the atom in question. Double and triple bonds count as one region of electron density. Lone pairs also count as one region of electron density.

This compound contains a double bond and two single bonds on each C atom.

There are three regions of electron density around each C atom. The double bond is one region.

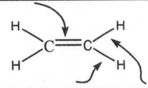

The single bonds are regions two and three.

Just as for BH_3, there are three regions of high electron density surrounding the C atom and the atom is sp^2 hybridized.

One π bond

Double bonds are made of one σ and one π bond.

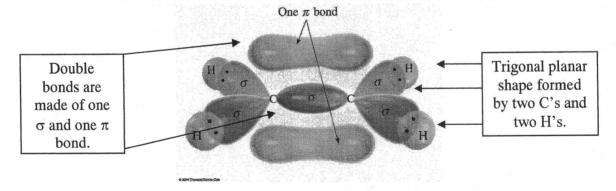

Trigonal planar shape formed by two C's and two H's.

51

This compound contains a triple bond and one single bond on each C atom.

$$H \colon C \colon\colon\colon C \colon H$$

There are two regions of electron density around each C atom. The triple bond is one region.

$$H \longleftarrow C \equiv C \longrightarrow H$$

The single bond is region two.

Just as for BeH_2, two regions of high electron density surround the C atom and the atom is sp^2 hybridized.

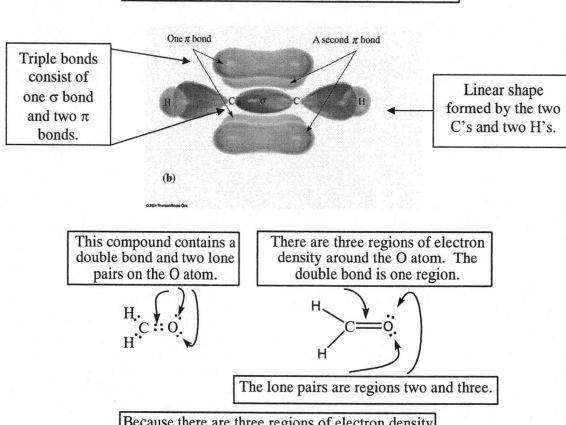

One π bond   A second π bond

Triple bonds consist of one σ bond and two π bonds.

Linear shape formed by the two C's and two H's.

(b)

© 2004 Thomson/Brooks Cole

This compound contains a double bond and two lone pairs on the O atom.

There are three regions of electron density around the O atom. The double bond is one region.

The lone pairs are regions two and three.

Because there are three regions of electron density around the O atom, the hybridization is sp^2.

52

3. **What is the hybridization of each of the indicated atoms in the amino acid alanine?**

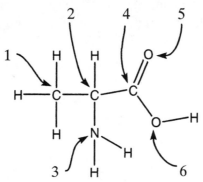

The correct answer is atom 1 is sp^3 hybridized, atom 2 is sp^3 hybridized, atom 3 is sp^3 hybridized, atom 4 is sp^2 hybridized, atom 5 is sp^2 hybridized, and atom 6 is sp^3 hybridized.

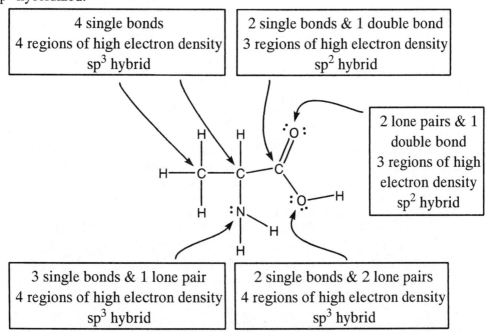

4 single bonds	2 single bonds & 1 double bond
4 regions of high electron density	3 regions of high electron density
sp^3 hybrid	sp^2 hybrid

2 lone pairs & 1 double bond
3 regions of high electron density
sp^2 hybrid

3 single bonds & 1 lone pair	2 single bonds & 2 lone pairs
4 regions of high electron density	4 regions of high electron density
sp^3 hybrid	sp^3 hybrid

INSIGHT: You must include the lone pairs to correctly answer this question. See Modules 9 and 10 for help determining lone pairs of electrons.

Polarity of Molecules Sample Exercise

4. **Which of the following molecules are polar?**

BH$_3$, CH$_2$F$_2$, H$_2$O, SF$_6$

The correct answer is CH$_2$F$_2$ and OCl$_2$ are polar. BH$_3$ and SF$_6$ are nonpolar.

To be polar molecules must have two essential features.
1) The molecule must contain at least one polar bond or one lone pair of electrons.   2) The molecule must be asymmetrical so that the bond dipoles do not cancel each other.

BH₃ contains polar bonds but the B-H bonds are symmetrical. Thus the dipoles for the polar bonds cancel each other and the molecule is nonpolar.

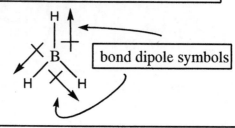

bond dipole symbols

CH_2F_2 contains 4 polar bonds. The two C-H bonds have their bond dipole pointed toward the C atom (C is more electronegative than H). The two C-F bonds have bond dipoles that are pointed away from the C atom (F is more electronegative than C). The result is an asymmetrical charge distribution making the molecule polar.

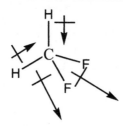

INSIGHT: Because CH_2F_2 is tetrahedral, every possible arrangement of the atoms is polar.

H_2O has two polar bonds and two lone pairs. Both bond dipoles for the O-H bonds are directed toward the O atom (O is more electronegative than H). These reinforce the large negative effect of the lone pairs making H_2O quite polar.

SF_6 has six polar S-F bonds. Bond dipoles are all directed toward the F atoms. But because this is a symmetrical arrangement of the dipoles, they cancel each other out leaving a nonpolar molecule.

Module 12
Acids and Bases

Introduction
There are three common theories of acids and bases that are commonly discussed in general chemistry texts. These are the Arrhenius, Brønsted-Lowry, and Lewis acid-base theories. In this module the important concepts to understand are <u>the distinctions and the commonalities between the three theories plus how to distinguish between compounds that act as acids or bases in one theory but not in another.</u>

Module 12 Key Equations & Concepts

1. **Acids produce H^+ in aqueous solution; bases produce OH^- in aqueous solution.**
   Arrhenius acid-base theory is the most restrictive theory requiring that the compound have an H^+ or OH^- and that it be dissolved in water.

2. **Acids are proton donors; bases are proton acceptors.**
   Brønsted-Lowry acid-base theory removes the restrictions that bases contain OH^- ions and that the solutions must be aqueous.

3. **Acids are electron pair acceptors; bases are electron pair donors.**
   Lewis acid-base theory is the least restrictive theory. The restriction on protons (H^+) being in the acid and all restrictions on solutions are removed.

Arrhenius Acid-Base Theory Sample Exercise
1. *Which of these compounds are Arrhenius acids and bases?*
   $$HCl, NaOH, H_2SO_4, BCl_3, Na_2CO_3, Ba(OH)_2, C_2H_4$$
   The correct answer is HCl and H_2SO_4 are Arrhenius acids. NaOH, $Ba(OH)_2$, and KOH are all Arrhenius bases.

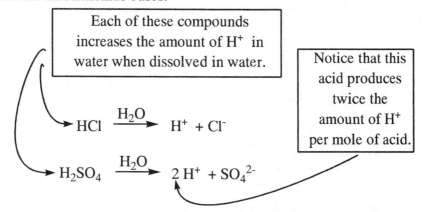

INSIGHT: To identify Arrhenius acids look for compounds that dissociate or ionize in water forming H^+.

55

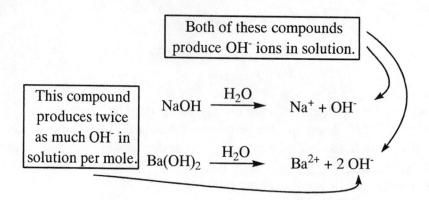

Both of these compounds produce OH⁻ ions in solution.

This compound produces twice as much OH⁻ in solution per mole.

$$NaOH \xrightarrow{H_2O} Na^+ + OH^-$$

$$Ba(OH)_2 \xrightarrow{H_2O} Ba^{2+} + 2\,OH^-$$

INSIGHT: To identify Arrhenius bases look for compounds that dissociate or ionize in water producing OH⁻.

YIELD BF_3 and Na_2CO_3 are not Arrhenius acids or bases because there are no acidic H's or basic OH's. C_2H_4 has H's but they are not acidic because the C-H bond is too strong to be easily broken.

Brønsted-Lowry Acid-Base Theory Sample Exercise

2. *Which of these compounds can be classified as Brønsted-Lowry acids and bases?*

$$HF,\ NH_3,\ H_2SO_4,\ BCl_3,\ Na_2CO_3,\ K_2S$$

The correct answer is HF and H_2SO_4, are Brønsted-Lowry acids. NH_3, Na_2CO_3, and K_2S are Brønsted-Lowry bases.

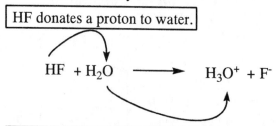

HF donates a proton to water.

$$HF + H_2O \longrightarrow H_3O^+ + F^-$$

The donated proton combines with H_2O to make H_3O^+.

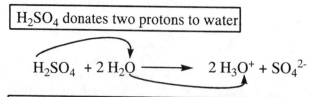

H_2SO_4 donates two protons to water

$$H_2SO_4 + 2\,H_2O \longrightarrow 2\,H_3O^+ + SO_4^{2-}$$

The donated protons combine with H_2O to make $2\,H_3O^+$.

NH₃ accepts a proton, H⁺, from H₂O.

$$NH_3 + H_2O \longrightarrow NH_4^+ + OH^-$$

The proton combines with NH₃ to form NH₄⁺.

NH₃ accepts a proton, H⁺, from HCl.

$$NH_3 + HCl \longrightarrow NH_4Cl_{(s)}$$

The proton combines with NH₃ to form NH₄⁺. This reaction is an example of a <u>nonaqueous</u> Brønsted-Lowry acid-base reaction.

Carbonate ion, CO₃²⁻, accepts a proton from water.

$$CO_3^{2-} + H_2O \longrightarrow HCO_3^- + OH^-$$

The proton combines with CO₃²⁻ to form HCO₃⁻
The Na⁺ ions are spectator ions in this reaction.

INSIGHT: | **Anions of weak acids will always be Brønsted-Lowry bases.** The carbonate ion, CO_3^{2-}, is the anion of the weak acid carbonic acid, H_2CO_3.

YIELD | Because Arrhenius and Brønsted-Lowry acids are both proton donors, they can both be identified similarly. Brønsted-Lowry bases however may not contain hydroxide ions, OH⁻. Instead they will accept a proton from water and form hydroxide ions in aqueous solutions.

3. *Identify the Brønsted-Lowry acid-base conjugate pairs in these reactions.*

$$CH_3COOH + H_2O \rightleftarrows CH_3COO^- + H_3O^+$$

$$F^- + H_2O \rightleftarrows HF + OH^-$$

The correct answer is CH₃COOH is an acid and its conjugate base is CH₃COO⁻. H₂O is a base and its conjugate acid is H₃O⁺ in the first reaction. In the second reaction F⁻ is a base and its conjugate acid is HF. H₂O is an acid and its conjugate base is OH⁻.

CH₃COOH donates a proton to H₂O making it an acid. It's conjugate base, CH₃COO⁻, differs from the acid by the loss of a single proton, H⁺.

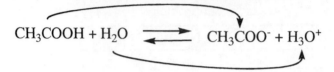

$$CH_3COOH + H_2O \rightleftharpoons CH_3COO^- + H_3O^+$$

H₂O accepts a proton from CH₃COOH making it a base. It's conjugate acid, H₃O⁺, differs from the base by the addition of a single proton, H⁺.

INSIGHT: | Many textbooks use this symbolism to designate the acid-base pairs. The 1's indicate that CH₃COOH and CH₃COO⁻ form one acid-base conjugate pair and the 2's indicate the members of the second acid-base pair.

$$CH_3COOH + H_2O \rightleftharpoons CH_3COO^- + H_3O^+$$

$$\text{acid}_1 \qquad \text{base}_2 \qquad\qquad \text{base}_1 \qquad \text{acid}_2$$

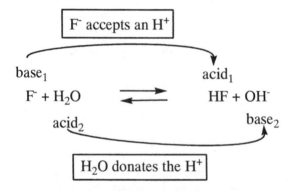

F⁻ accepts an H⁺

$$\text{base}_1 \qquad\qquad\qquad \text{acid}_1$$
$$F^- + H_2O \rightleftharpoons HF + OH^-$$
$$\text{acid}_2 \qquad\qquad\qquad \text{base}_2$$

H₂O donates the H⁺

YIELD

Notice that in one reaction H₂O is a base and in the other reaction it is an acid. **In Brønsted-Lowry theory all acid-base reactions are a competition between stronger and weaker acids or bases.** In the CH₃COOH reaction, the stronger acid is CH₃COOH, thus water acts as a base in its presence. In the F⁻ reaction, H₂O is the stronger acid so it acts as the acid in this reaction. Water is an amphoteric species; it can be either an acid or a base in the presence of a stronger acid or base.

4. *Arrange the following species in order of their strength as bases.*

$$HCO_3^-, Cl^-, CO_3^{2-}$$

The correct answer is $Cl^- < HCO_3^- < CO_3^{2-}$.

INSIGHT:

> The best approach to a problem like this is to recognize which acid or base the species are derived from. The Cl^- ion is derived from the acid HCl, HCO_3^- derives from H_2CO_3, and CO_3^{2-} derives from HCO_3^-. (In each case the acid the species derives from is determined by adding H^+ to the species. In other words, we have determined the conjugate acids of each species.) Now we can easily compare the strengths of the conjugate acids and determine the base strengths. For acid strengths, HCl is by far the strongest acid, H_2CO_3 is the next strongest acid, and finally the HCO_3^- ion is the weakest acid. In fact, the HCO_3^- ion is a weak base.

YIELD

> Some important things to remember about acids and bases are:
> 1) **The stronger the acid, the weaker the conjugate base.**
> 2) **The weaker the acid, the stronger the conjugate base.**
> 3) **The stronger the base, the weaker the conjugate acid.**
> 4) **The weaker the base, the stronger the conjugate acid.**

INSIGHT:

> Since the Cl^- ion is the conjugate base of HCl, a strong acid, it is the weakest base. The HCO_3^- ion is the conjugate base of the weak acid H_2CO_3, thus it is a stronger base than Cl^-. Finally, the CO_3^{2-} ion is the conjugate base of the very weak acid (a basic compound is a very weak acid) HCO_3^- making it the strongest base.

Lewis Acid-Base Theory Sample Exercises

5. *Identify the Lewis acids and bases in each of these reactions.*

$$NH_3 + HCl \rightarrow NH_4Cl$$
$$BCl_3 + NH_3 \rightarrow BCl_3NH_3$$

The correct answer is HCl and BCl_3 are the Lewis acids. NH_3 is the Lewis base in both reactions.

YIELD

> The only way that Lewis acids and bases can be determined is to draw the Lewis dot structures and determine what is happening to the lone pairs of electrons. It is also important to realize that the other two acid-base theories are focused on the protons, H^+. Lewis acid-base theory focuses on electrons. Consequently, the actions are reversed. <u>Acids are *donors*</u> in Arrhenius and Brønsted-Lowry theories. **<u>In Lewis theory acids are electron pair *acceptors*</u>.**

This lone pair of electrons is donated to the H$^+$ from HCl to form NH$_4^+$.

This lone pair of electrons is donated to the B in BCl$_3$ to form NH$_3$BCl$_3$.

Module 13
States of Matter

Introduction

This module describes the basic laws that govern the three states of matter: gas, liquid, and solid. In this module the most important things to understand are <u>how to use the combined and ideal gas laws to calculate various quantities of gases; Graham's law of effusion to determine the effusion rates of various gases; how to determine the relative freezing and boiling points of various liquids based on their intermolecular forces; how to determine the relative melting points of various solids based on the strength of their bonding; and how to use the number of particles in the three cubic unit cells to calculate the radii of atoms.</u>

Module 13 Key Equations & Concepts

1. $\dfrac{P_1 V_1}{T_1} = \dfrac{P_2 V_2}{T_2}$

   The combined gas law is a combination of Boyle's and Charles's gas laws. It is used to determine a new temperature, volume, or pressure of a gas given the original temperature, volume and pressure.

2. $PV = nRT$

   The ideal gas law is used to calculate any one of these quantities, the pressure, volume, temperature, or number of moles of a gas given three of the other quantities. This equation is often used in reaction stoichiometry problems involving gases.

3. $\dfrac{R_1}{R_2} = \sqrt{\dfrac{M_2}{M_1}}$

   Graham's law of effusion is used to determine how quickly one gas effuses (or diffuses) relative to another gas. It can also be used to determine the molar masses of gases based on their effusion rates.

4. **Ion-Ion interactions, Dipole-Dipole interactions, Hydrogen bonding, London Dispersion forces**

   These are the four basic intermolecular forces involved in liquids. The strength of these interactions will determine the boiling points of each liquid.

5. **Simple Cubic Unit Cells contain 1 particle per unit cell**

   The simplest type of cubic unit cell has an atom, ion, or molecule at the each of the corners. Because the atoms, ions, and molecules are shared from unit cell to unit cell, each one contributes one-eighth of its volume to a single unit cell. Thus there are 8 x 1/8 = 1 atom, ion, or molecule per unit cell.

6. **Body-centered Cubic Unit Cells contain 2 particles per unit cell**

   Body-centered cubic unit cells have one more atom, ion, or molecule in the center of the unit cell. Thus there are 8 x 1/8 + 1 = 2 atoms, ions, or molecules per unit cell.

---

7. **Face-centered Cubic Unit Cells contain 4 particles per unit cell**

Face-centered cubic unit cells have six additional atoms, ions, and molecules (one in each face of the cube.) These atoms, ions, and molecules are shared one-half in each unit cell. Thus there are (8 x 1/8) + (6 x 1/2) = 4 atoms, ions, or molecules per unit cell.

8. **Covalent Network Solids, Ionic solids, Metallic solids, Molecular solids**

These are the four basic types of solids. The strength of the bonds in solids determines the freezing and boiling points of each.

Gas Law Sample Exercises

1. *A sample of a gas initially having a pressure of 1.25 atm and volume of 3.50 L has its volume changed to 7.50 x 10⁴ mL at constant temperature. What is the new pressure of the gas sample?*

The correct answer is 0.0583 atm.

$$7.50 \times 10^4 \text{ mL}\left(\frac{1 \text{ L}}{1000 \text{ mL}}\right) = 75.0 \text{ L}$$

$\dfrac{P_1 V_1}{T_1} = \dfrac{P_2 V_2}{T_2}$ simplifies to $P_1 V_1 = P_2 V_2$ at constant temperature $(T_1 = T_2)$

$$1.25 \text{ atm} \times 3.50 \text{ L} = P_2 \times 75.0 \text{ L}$$

$$\frac{1.25 \text{ atm} \times 3.50 \text{ L}}{75.0 \text{ L}} = P_2$$

$$0.0583 \text{ atm} = P_2$$

INSIGHT: It is very important in these problems that the proper units be used. In this problem we must change the mL to L or vice versa. Also, because the problem is at constant temperature, the combined gas law can be simplified to Boyle's law.

2. *A gas sample initially having a pressure of 1.75 atm, and a volume of 4.50 L at a temperature of 25.0°C is heated to 37.0°C at a pressure of 1.50 atm. What is the gas's new volume?*

The correct answer is 5.46 L.

$$\frac{P_1 V_1}{T_1} = \frac{P_2 V_2}{T_2} \text{ where :}$$

$$P_1 = 1.75 \text{ atm}, V_1 = 4.50 \text{ L}, T_1 = 25.0°\text{C} = 298.1 \text{ K}$$

$$P_2 = 1.50 \text{ atm and } T_2 = 37.0°\text{C} = 310.1 \text{ K}$$

$$V_2 = \frac{P_1 V_1 T_2}{T_1 P_2} = \frac{(1.75 \text{ atm})(4.50 \text{ L})(310.1 \text{ K})}{(298.1 \text{ K})(1.50 \text{ atm})} = 5.46 \text{ L}$$

 YIELD All gas law problems involving temperature must be in units of Kelvin. Be absolutely certain that you convert temperatures into Kelvin when working any gas law problems.

3. **A gas sample at a pressure of 3.50 atm and a temperature of 45.0°C has a volume of 1.65 x 10³ mL. How many moles of gas are in this sample?**

The correct answer is 0.221 moles.

$$PV = nRT \text{ where :}$$

$$P = 3.50 \text{ atm}, V = 1.65 \times 10^3 \text{ mL} = 1.65 \text{ L}, R = 0.0821 \frac{\text{L atm}}{\text{mol K}}, T = 45.0°C = 318.1 \text{ K}$$

$$n = \frac{PV}{RT} = \frac{(3.50 \text{ atm})(1.65 \text{ L})}{\left(0.0821 \frac{\text{L atm}}{\text{mol K}}\right)(318.1 \text{ K})} = 0.221 \text{ mol}$$

INSIGHT: R is the ideal gas constant. In gas laws, it will have the following value and units, R = 0.0821 L atm/mol K. This defines the units that we must use in the ideal gas law. P must be in atm, V in L, n in moles, and T in K.

4. **How many grams of $CO_{2(g)}$ are present in 11.2 L of the gas at STP?**

The correct answer is 22.0 g.

INSIGHT: STP is a symbol for standard temperature and pressure. When you see those symbols in a problem involving gases, you may assume that the temperature is 0.00° C or 273.15 K and the pressure is 1.00 atm or 760 mm Hg.

$$PV = nRT \text{ thus } n = \frac{PV}{RT}$$

$$n = \frac{(1.00 \text{ atm})(11.2 \text{ L})}{\left(0.0821 \frac{\text{L atm}}{\text{mol K}}\right)(273.15 \text{ K})} = 0.500 \text{ mol}$$

$$0.500 \text{ mol} \left(\frac{44.0 \text{ g CO}_2}{1 \text{ mol CO}_2}\right) = 22.0 \text{ g CO}_2$$

5. **If 35.0 g of Al are reacted with excess sulfuric acid, how many L of hydrogen gas, H_2, will be formed at 1.25 atm of pressure and 75.0°C?**

$$2 Al_{(s)} + 3 H_2SO_{4(aq)} \rightarrow Al_2(SO_4)_{3(aq)} + 3 H_{2(g)}$$

The correct answer is 44.6 L.

a) Calculate the number of moles of hydrogen gas formed in the reaction.

$$(35.0 \text{ g Al})\left(\frac{1 \text{ mol}}{26.98 \text{ g Al}}\right)\left(\frac{3 \text{ mol H}_2}{2 \text{ mol Al}}\right) = 1.95 \text{ mol}$$

b) Use the ideal gas law to determine the volume of the gas.

Remember that we must convert 75.0° C to 348.1 K.

$$PV = nRT \text{ thus } V = \frac{nRT}{P}$$

$$V = \frac{(1.95 \text{ mol})\left(0.0821 \frac{\text{L atm}}{\text{mol K}}\right)(348.1 \text{ K})}{1.25 \text{ atm}} = 44.6 \text{ L}$$

INSIGHT: | Sample Exercise 5 uses a combination of reaction stoichiometry and the ideal gas law to determine the volume of the gas formed in a reaction.

6. A gas having a molar mass of 16.0 g/mol effuses through a pinhole 4.00 times faster than an unknown gas. What is the molar mass of the unknown gas?
The correct answer is 256 g/mol.

Graham's law of effusion relates the rate at which molecules effuse to the molar masses of the substances. In this case R_1 = effusion rate of gas 1, R_2 = effusion rate of gas 2, M_1 = molar mass of gas 1, M_2 = molar mass of gas 2. In this problem $R_1 = 4.00$, $R_2 = 1.00$, $M_1 = 16.0$, and M_2 is unknown.

$$\frac{R_1}{R_2} = \sqrt{\frac{M_2}{M_1}} \text{ thus } \frac{4.00}{1.00} = \sqrt{\frac{M_2}{16.0 \text{ g/mol}}} \text{ to find } M_2 \text{ square both sides of this equation}$$

$$16.0 = \frac{M_2}{16.0 \text{ g/mol}} \text{ thus } M_2 = 16.0 \times 16.0 \text{ g/mol} = 256 \text{ g/mol}$$

INSIGHT: | Deciding which gas has the faster rate can be confusing. However, as long as you associate the rate with one of the gases, i.e. R_1 with M_1 or R_2 with M_2, the relationship will work correctly.

Liquid Sample Exercise

7. Arrange these substances based on their boiling points from lowest to highest.
CO_2, $NaCl$, C_2H_5OH, CH_3Cl
The correct answer is $CO_2 < CH_3Cl < C_2H_5OH < NaCl$.

Boiling points are determined by the strength of the intermolecular forces present in a liquid. In general, ion-ion interactions are strongest, hydrogen bonding is next, dipole-dipole interactions are third strongest, and the weakest intermolecular force is London dispersion forces. The strongest intermolecular forces in liquid CO_2 are London dispersion forces, CH_3Cl's strongest intermolecular forces are dipole-dipole interactions, hydrogen bonding is dominant in C_2H_5OH, and $NaCl$ is an ionic compound. Thus the correct order is $CO_2 < CH_3Cl < C_2H_5OH < NaCl$.

YIELD | The strongest intermolecular forces in *nonpolar molecules are London dispersion forces. Polar molecules have dipole-dipole interactions as their* dominant intermolecular force. *Molecules that have one or more H atoms directly bonded to a N, O, or F atom have hydrogen bonding as their* dominant intermolecular force. *Ion-ion interactions are found in ionic compounds.* Review Module 11 for rules on molecular polarity.

Solid Sample Exercises

8. **Lead, Pb, has a density of 11.35 g/cm³. Solid Pb crystallizes in one of the cubic unit cells (either simple, body-centered, or face-centered cubic) which has an edge length of 4.95 x 10⁻⁸ cm. Which of the three unit cells is present in solid Pb? What is the radius, in cm, of a Pb atom?**

The correct answer is face-centered cubic and the radius is 1.75 x 10⁻⁸ cm.

a) First we need to determine the volume of a single unit cell.

For cubes $V = \ell^3$.

$$V = (4.95 \times 10^{-8}\ cm)^3 = 1.21 \times 10^{-22}\ cm^3$$

b) Use the volume and the density to determine the mass of a single unit cell.

$$?\ g = 1.21 \times 10^{-22}\ cm^3 \left(\frac{11.35\ g}{cm^3} \right) = 1.37 \times 10^{-21}\ g$$

c) Determine the mass of a single Pb atom.

$$?\ g = 207.2\ g/mol \left(\frac{1\ mol}{6.022 \times 10^{23}\ atoms} \right) = 3.44 \times 10^{-22}\ g/atom$$

d) Use steps b and c to determine the number of atoms in a single unit cell.

$$?\ atoms = \frac{1.37 \times 10^{-21}\ g}{3.44 \times 10^{-22}\ g/atom} = 3.98\ atoms \approx 4\ atoms\ in\ the\ unit\ cell$$

Because there are 4 atoms in this unit cell we can conclude it is a face - centered cubic unit cell.

YIELD

The fact that *simple cubic unit cells contain one particle* per unit cell, *body-centered cubic unit cells contain two*, and *face-centered cubic unit cells contain four* is one of the key pieces of information that you will need to work these cubic unit cell problems.

e) To calculate the radius of a single Pb atom requires use of the Pythagorean theorem and some knowledge of the geometry of a face - centered cubic unit cell.

(ii)

Notice that in this picture of one face of a face-centered unit cell that four atoms form the face diagonal. The face diagonal length is $\sqrt{2}$ times the edge length and that there are 4 atomic radii contained in the face diagonal.

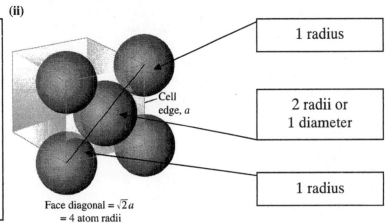

1 radius

2 radii or 1 diameter

1 radius

Cell edge, a

Face diagonal = $\sqrt{2} a$
= 4 atom radii

$$\text{diagonal length} = \sqrt{2}\left(4.95 \times 10^{-8} \text{ cm}\right) = 7.00 \times 10^{-8} \text{ cm}$$

$$\text{radius of a Pb atom} = \frac{7.00 \times 10^{-8} \text{ cm}}{4 \text{ radii}} = 1.75 \times 10^{-8} \text{ cm}$$

INSIGHT:

For the other cubic unit cells the geometrical relationships are:
1) **simple cubic unit cells** - atomic radius = ½ cell edge length
2) **body-centered cubic unit cells** – atomic radius = $\sqrt{3}$ x ¼ cell edge length

8. *Arrange these substances based on their melting points from lowest to highest.*
CO_2, KCl, Na, SiO_2

The correct answer is $CO_2 <$ Na $<$ KCl $< SiO_2$.

INSIGHT:

Melting points of solids are determined by the strength of the forces bonding them together. In general, the weakest forces are intermolecular forces found in molecular solids like CO_2, next weakest are metallic bonds as in Na, ionic bonds are relatively strong like in KCl, and the strongest are the covalent bonds from atom to atom that bond network covalent species like SiO_2.

YIELD

The real key to these melting point problems is determining a solid's classification.
1) **Molecular solids** will always be covalent compounds that form individual molecules. Most of the covalent species that you have learned up to now will be molecular solids.
2) **Metallic solids** are by far the easiest to classify. Look for a metallic element.
3) **Ionic solids** are the basic ionic compounds that you have learned up to this point.
4) The hardest substances to classify are the **network covalent species**. They are covalent species that form extremely large molecules through extended arrays of atoms that are covalently bonded. Most textbooks have a list of the common network covalent solids which include diamond, graphite, tungsten carbide (WC), and sand (SiO_2). It is probably best that you know these molecules by heart.

Module 14
Solutions

Introduction

This module discusses the properties of solutions. The most important things to understand from the module are how to: <u>predict which species will dissolve in a given solvent based on molecular polarity; increase the solubility of a given species in a solvent; convert from one concentration unit to any of the other three units; use Raoult's law to predict the vapor pressure of a solution; determine the freezing and boiling points of solutions; and calculate the osmotic pressure of a solution.</u>

Module 14 Key Equations & Concepts

1. **Like Dissolves Like**

   This rule is a statement of the common phenomenon that polar molecules are readily soluble in other polar molecules and that nonpolar molecules are readily soluble in other nonpolar molecule. However, polar molecules are fairly insoluble in nonpolar molecules.

2. **Solute solubility is increased when the solvent is *heated* in an *endothermic* dissolution. Solute solubility is increased when the solvent is *cooled* in an *exothermic* dissolution. Increasing a gas's pressure will increase the solubility of a gas in a liquid.**

   These are the basic rules for changing the solubility of various substances in solution. They are used to determine methods to increase solubilities of substances.

3.
$$M = \frac{\text{moles of solute}}{\text{L of solution}}$$

$$m = \frac{\text{moles of solute}}{\text{kg of solvent}}$$

$$\% \text{ w/w} = \frac{\text{mass of one solution component}}{\text{mass of total solution}} \times 100$$

$$X_A = \frac{\text{moles of component A}}{\text{total moles of solution}}$$

   The four most common units of solution concentration are molarity (M) used in reaction stoichiometry and osmotic pressure problems, molality (m) used in the freezing point depression and boiling point elevation relationships, percent by mass (% w/w) used frequently for concentrated solutions, and mole fraction (X_A) used in Raoult's law. The definitions of each concentration unit are given above.

4. $P_{\text{solvent}} = X_{\text{solvent}} P^0_{\text{solvent}}$

   Raoult's law is used to determine the vapor pressure of a solution containing a nonvolatile solute.

---

5. $\Delta T_f = iK_f m$ and $\Delta T_b = iK_b m$

   The freezing point depression and boiling point elevation relationships describe how much the freezing or boiling temperatures of a solution will differ from the pure solvent's freezing and boiling points.

6. $\Pi = MRT$

   The osmotic pressure of solutions can be easily calculated using this relationship.

Solubility of a Solute in a Given Solvent Sample Exercise

1. **Which of the following substances are soluble in water?**

   $SiCl_4$, NH_3, C_8H_{18}, $CaCl_2$, CH_3OH, $Ca_3(PO_4)_2$

   The correct answer is NH_3, $CaCl_2$, and CH_3OH are soluble in water. The others are insoluble in water.

The "Like Dissolves Like" rule implies that polar species dissolve in polar species and nonpolar species dissolve in nonpolar species. Consequently, nonpolar species do not dissolve in polar species and polar species do not dissolve in nonpolar species. In this example, NH_3 and CH_3OH are both polar covalent compounds so they will dissolve in the highly polar solvent water. $CaCl_2$ is an ionic compound which is water soluble (the solubility rules also apply in these problems). $SiCl_4$ and C_8H_{18} are both nonpolar covalent compounds thus they are insoluble in water. $Ca_3(PO_4)_2$ is an ionic compound that is insoluble in water. Refresh your memory of the solubility rules if necessary.

In these problems there are two important considerations to have in mind. 1) Are the covalent compounds in the problem polar or nonpolar? 2) Are the ionic compounds in the problem soluble or insoluble based on the solubility rules? Remember the strong acids and bases from previous modules will also be water soluble.

Increasing the Solubility of a Solute in a Given Solvent Sample Exercises

2. **Given the following dissolution in water equation, which of these changes in conditions are correct statements?**

$$KI_{(s)} \xrightarrow{\ H_2O\ } K^+_{(aq)} + I^-_{(aq)} \quad \Delta H_{dissolution} > 0$$

   a) *Increasing the temperature of the solvent will increase the solubility of the compound in the solvent.*

   b) *Decreasing the temperature of the solvent will increase the solubility of the compound in the solvent.*

   c) *Changing the temperature of the solvent will not affect the solubility of the compound in the solvent.*

   d) *Increasing the pressure of the solute will increase the solubility of the compound in the solvent.*

   e) *Increasing the pressure of the solute will decrease the solubility of the compound in the solvent.*

   f) *Increasing the pressure of the solute will not affect the solubility of the compound in the solvent.*

68

The correct answer is conditions a) and f) are correct and no others are correct.

> There are several important hints in this problem to help you answer it. The $KI_{(s)}$ indicates that this is a solid dissolving in water. Changing the pressure of liquids and solids have no effect on their solubilities. The positive $\Delta H_{dissolution}$ indicates that this is an <u>endothermic</u> process. Heating the solvent for endothermic dissolutions increases the solubility of the solute.
>
> **$\Delta H_{dissolution} < 0$ is exothermic. $\Delta H_{dissolution} > 0$ is endothermic.**

3. ***Given the following dissolution in water equation, which of these changes in conditions are correct statements?***

$$O_{2(g)} \xrightarrow{\;H_2O\;} O_{2(aq)} \quad \Delta H_{dissolution} < 0$$

  a) ***<u>Increasing</u> the temperature of the solvent will <u>increase</u> the solubility of the compound in the solvent.***

  b) ***<u>Decreasing</u> the temperature of the solvent will <u>increase</u> the solubility of the compound in the solvent.***

  c) ***Changing the temperature of the solvent will <u>not affect</u> the solubility of the compound in the solvent.***

  d) ***<u>Increasing</u> the pressure of the solute will <u>increase</u> the solubility of the compound in the solvent.***

  e) ***<u>Increasing</u> the pressure of the solute will <u>decrease</u> the solubility of the compound in the solvent.***

  f) ***<u>Increasing</u> the pressure of the solute will <u>not affect</u> the solubility of the compound in the solvent.***

The correct answer is conditions b) and d) are correct and no others are correct.

> The important hints in this problem are 1) $O_{2(g)}$ indicates that this is a gas dissolving in water. Increasing the pressure of gases has a significant effect on their solubilities. In general, increasing the pressure of a gas will increase its solubility in a liquid. 2) The negative $\Delta H_{dissolution}$ indicates that this is an <u>exothermic</u> process. Heating the solvent for exothermic dissolutions decreases the solubility of the solute. Cooling the solvent increases the solubility of the solute in exothermic dissolutions.

INSIGHT: 1) Pay attention to whether the substance being dissolved is a solid, liquid, or gas. That will tell you if the changing pressure condition is applicable. 2) Pay attention to whether or not the dissolution is endo- or exothermic. That is your hint as to heating or cooling the solvent will increase the solubility of the substance. Both of these effects are ramifications of LeChatelier's principle.

Concentration Unit Conversion Sample Exercises

4. ***An aqueous sulfuric acid solution that is 3.75 M has a density of 1.225 g/mL. What is the concentration of this solution in molality (m), percent by mass (% w/w) of H_2SO_4, and mole fraction ($X_{sulfuric\ acid}$) of H_2SO_4?***
    The correct answer is 4.38 *m*, 30.0 % w/w, and $X_{sulfuric\ acid} = 0.0730$.

$$3.75\ M\ H_2SO_4 = \frac{3.75\ \text{moles of }H_2SO_4}{1.00\ \text{L of solution}}$$

We must separate the masses of the solute and solvent to determine the other concentrations. Molarity tells us the volume of the solution, not the mass. The solution's density will help us calculate the mass of this solution. To make the calculation as easy as possible, we can assume that that we have one liter of this 3.75 M solution.

a) $1.00\ \text{L of }3.75\ M\ H_2SO_4 = 1000\ \text{mL}\left(\dfrac{1.225\ g}{mL}\right) = 1225\ \text{g of }3.75\ M\ H_2SO_4$ — mass of the solution

b) $3.75\ \text{mole }H_2SO_4\left(\dfrac{98.1\ g\ H_2SO_4}{1\ \text{mole }H_2SO_4}\right) = 368\ \text{g }H_2SO_4$ — mass of the solute

c) $1225\ g - 368\ g = 857\ g$ or $0.857\ \text{kg of water, the solvent}$ — mass of the solvent

d) $m = \dfrac{\text{moles of solute}}{\text{kg of solvent}} = \dfrac{3.75\ \text{moles of }H_2SO_4}{0.857\ \text{kg of }H_2O} = 4.38\ m\ H_2SO_4$

e) $\%\ w/w = \dfrac{\text{mass of }H_2SO_4}{\text{mass of solution}} \times 100 = \dfrac{368\ \text{g of }H_2SO_4}{1225\ \text{g of solution}} \times 100 = 30.0\%\ H_2SO_4$

f) $\text{moles of }H_2O = 857\ g\left(\dfrac{1\ \text{mole of }H_2O}{18\ \text{g of }H_2O}\right) = 47.6\ \text{moles of }H_2O$ — moles of solvent for the mole fraction

g) $X_{\text{sulfuric acid}} = \dfrac{\text{moles of solute}}{\text{moles of solute} + \text{moles of solution}} = \dfrac{3.75\ \text{moles}}{3.75\ \text{moles} + 47.6\ \text{moles}} = 0.0730$

YIELD — Converting solution concentrations from molarity to the other three concentration units is by far the hardest type of these conversion problems. The key to doing this correctly is separating the masses of the solute and solvent from the mass of the solution.

5. *An aqueous sucrose, $C_{12}H_{22}O_{11}$, solution that is 11.0 % w/w has a density of 1.0432 g/mL. What is the concentration of this solution in molarity (M), molality (m), and mole fraction ($X_{sucrose}$) of $C_{12}H_{22}O_{11}$?*

The correct answer is 0.335 *M*, 0.361 *m*, and 0.00646 X_{sucrose}.

If we assume that we have 100.0 g of solution, we can conclude that we have 11.0 g of sucrose and 89.0 g of water. This is the key to solving this problem.

a) $11.0\ \text{g of }C_{12}H_{22}O_{11}\left(\dfrac{1\ \text{mole of }C_{12}H_{22}O_{11}}{342.3\ \text{g of }C_{12}H_{22}O_{11}}\right) = 0.0321\ \text{moles of }C_{12}H_{22}O_{11}, \text{the solute}$

b) $89.0\ \text{g of }H_2O\left(\dfrac{1\ \text{mole of }H_2O}{18.0\ \text{g of }H_2O}\right) = 4.94\ \text{moles of }H_2O, \text{the solvent}$ — converting the solution mass to volume

c) $\text{volume of }100.0\ \text{g of this solution} = 100.0\ g\left(\dfrac{1.00\ mL}{1.0432\ g}\right) = 95.9\ \text{mL} = 0.0959\ \text{L}$

d) $M = \dfrac{\text{moles of sucrose}}{\text{L of solution}} = \dfrac{0.0321 \text{ moles of sucrose}}{0.0959 \text{ L of solution}} = 0.335\,M$

e) $m = \dfrac{\text{moles of sucrose}}{\text{kg of solvent}} = \dfrac{0.0321 \text{ moles of sucrose}}{0.0890 \text{ kg of water}} = 0.361\,m$

concentrations are easily determined from masses, moles, and volume

f) $X_{\text{sucrose}} = \dfrac{\text{moles of sucrose}}{\text{moles of sucrose} + \text{moles of water}} = \dfrac{0.0321 \text{ moles}}{0.0321 \text{ moles} + 4.94 \text{ moles}} = 0.00646$

INSIGHT: Notice that this problem is much easier because % w/w is a concentration unit that easily separates into the solute and the solvent.

Raoult's Law Sample Exercise

6. *What is the vapor pressure of a solution made by dissolving 11.0 g of sucrose in 89.0 g of water at 25.0°C? The vapor pressure of pure water at 25.0°C is 23.76 torr.*

   The correct answer is 23.60 torr.

   From exercise 5 shown above, we know that the $X_{\text{sucrose}} = 0.00646$. Thus the mole fraction of the solvent, water $= 1.00000 - 0.00646 = 0.99354$

$$P_{\text{solution}} = X_{\text{solvent}} P^0_{\text{solvent}}$$
$$P_{\text{solution}} = 0.99354\,(23.76 \text{ torr}) = 23.60 \text{ torr}$$

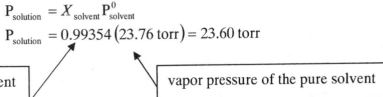

mole fraction of the solvent

vapor pressure of the pure solvent

Freezing Point Depression and Boiling Point Elevation Sample Exercises

7. *If 11.0 g of sucrose, a nonelectrolyte, are dissolved in 89.0 g of water, at what temperature will this solution boil under 1.00 atm of pressure? The boiling point elevation constant, K_b, for water is 0.512 °C/m.*

   The correct answer is 100.185 °C.

   This solution's concentration was determined in exercise 5, shown above, to be 0.361 *m*.

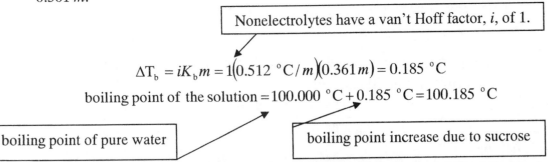

Nonelectrolytes have a van't Hoff factor, *i*, of 1.

$$\Delta T_b = iK_b m = 1(0.512\ ^\circ C/m)(0.361\,m) = 0.185\ ^\circ C$$

boiling point of the solution $= 100.000\ ^\circ C + 0.185\ ^\circ C = 100.185\ ^\circ C$

boiling point of pure water

boiling point increase due to sucrose

8. *12.4 g of a nonelectrolyte are dissolved in 100.0 g of water and the solution is then frozen. The freezing point of the solution is determined to be -5.00 °C. What is the molar mass of the nonelectrolyte? The freezing point depression constant, K_f, for water is 1.86 °C/m.*

   The correct answer is 46.1 g/mol.

$$\Delta T_f = iK_f m \text{ thus } m = \frac{\Delta T_f}{iK_f} = \frac{\Delta T_f}{K_f} \text{ for nonelectrolytes}$$

$$\Delta T_f = 0.00 \text{ °C} - (-5.00 \text{ °C}) = 5.00 \text{ °C}$$

$$m = \frac{5.00 \text{ °C}}{1.86 \text{ °C}/m} = 2.69 \text{ } m$$

$$2.69 \text{ } m = \frac{\text{moles of nonelectrolyte}}{\text{kg of solvent, water}} = \frac{? \text{ moles of nonelectrolyte}}{0.100 \text{ kg of water}}$$

$$? \text{ moles of nonelectrolyte} = 2.69 \text{ } m \text{ } (0.100 \text{ kg of water}) = 0.269 \text{ moles of nonelectrolyte}$$

$$\text{molar mass of nonelectrolyte} = \frac{\text{mass of nonelectrolyte}}{\text{moles of nonelectrolyte}} = \frac{12.4 \text{ g}}{0.269 \text{ mol}} = 46.1 \text{ g/mol}$$

9. *A 1.00 m solution of a strong` electrolyte dissolved in 100.0 g of water forms a solution having a freezing point of -5.58°C. Which of these generic ionic formulas would correspond to the formula of the electrolyte? M represents a typical metal cation and X a typical anion. The freezing point depression constant, K_f, for water is 1.86 °C/m.*

    a) *MX*
    b) *MX$_2$*
    c) *MX$_3$*
    d) *M$_2$X$_3$*
    e) *M$_2$X$_4$*

The correct answer is b) MX$_2$.

$$\Delta T_f = iK_f m \text{ which can be rearranged to } \frac{\Delta T_f}{K_f m} = i$$

$$\Delta T_f = 0.00 \text{ °C} - (-5.58 \text{ °C}) = 5.58 \text{ °C}$$

$$i = \frac{\Delta T_f}{K_f m} = \frac{5.58 \text{ °C}}{1.86 \text{ °C}/m \times 1.00 \text{ } m} = 3.00$$

The answer indicates that this electrolyte has 3 times the effect of a nonelectrolyte on the freezing point depression. Consequently, there must be 3 ions dissolved in solution. In the possible answers only MX$_2$ can dissolve to generate 3 ions in solution, namely 1 M ion and 2 X ions.

Osmotic Pressure Sample Exercise

10. *What is the osmotic pressure of a 0.335 M sucrose solution at 25.0°C?*
    The correct answer is 8.20 atm.

    $$\Pi = MRT \text{ where } \Pi \text{ is the osmotic pressure in atm,}$$

    $$M \text{ is the solution concentration in molarity,}$$

$$R \text{ is the ideal gas constant} = 0.0821 \text{ L atm}/\text{mol K}$$

$$\text{and T is the temperature in K.}$$

$$\Pi = 0.335 \text{ } M \left(0.0821 \text{ L atm}/\text{mol K}\right)(298.1 \text{ K})$$

$$\Pi = 8.20 \text{ atm}$$

Module 15
Heat Transfer, Calorimetry, and Thermodynamics

Introduction
This module is a brief discussion of heat related topics in chemistry. In this area we must address the following major issues: <u>the basic heat transfer equation and how that impacts both calorimetry and heating substances that remain in a single phase, simple chemical thermodynamics including the change in energy (ΔE) of a system, the heat (q) and work (w) involved in an energy change, the change in enthalpy (ΔH), calculation of ΔH using Hess's law, calculation of the change in entropy (ΔS), calculation of the Gibbs Free Energy change (ΔG) and the temperature dependence of the Gibbs Free Energy change.</u> All of these topics involve how heat and energy are transferred from one chemical system to another.

Module 15 Key Equations & Concepts

1. **$q = mC\Delta T$**

   This is the basic heat transfer equation which calculates the amount of energy emitted or absorbed when an object warms up or cools down. (q is the heat, m is the mass, C is the specific heat, and ΔT is the temperature change.) It is used in calorimetry, heat lost = heat gained problems, and to determine the heat necessary to either heat up or cool down a substance that remains in a single phase.

2. **$\Delta E = q + w$**

   The change in the energy of a chemical system is determined by two factors, 1) how much heat (q) enters or leaves the system and 2) how much work (w) the system does in the form of expanding or contracting against a constant pressure such as the atmosphere. The correct signs (i.e., +q or –q) of the heat and work are crucial to understanding these problems.

3. **$w = - P\Delta V = - \Delta n_{gas}RT$ at constant temperature and pressure**

   This relationship defines the amount of work that a system can do at constant temperature and pressure. (w is the work, P is the pressure, ΔV is the volume change, Δn_{gas} is the change in the number of moles of gas, R is the ideal gas constant, and T is the temperature.) It also describes the work that a system can do or have done on it when there is a change in the number of moles of gas.

4. **$\Delta H = \Delta E + P\Delta V = q_P$ at constant temperature and pressure**

   This is the definition of enthalpy, ΔH, and its relationship to the energy change of a system. (q_p is the heat flow at constant pressure.)

5. **$\Delta H^0_{rxn} = \sum n\Delta H^0_{f\ products} - \sum n\Delta H^0_{f\ reactants}$**

   This is one form of Hess's law which is used to determine the heat absorbed or released in a chemical reaction from the heats of formation of the products and reactants. ($\Delta H^0_{f\ products}$ is the heat of formation of the product substances at standard conditions. $\Delta H^0_{f\ reactants}$ is the heat of formation of the reactant substances at standard conditions. n represents the stoichiometric coefficients in the balanced chemical reaction.)

---

6. $\Delta S^0_{rxn} = \sum n S^0_{f\ products} - \sum n S^0_{f\ reactants}$

This is the relationship for determining the entropy change of a chemical reaction given the standard entropies of formation. ($\Delta S^0_{f\ products}$ is the entropy of formation of the product substances at standard conditions. $\Delta S^0_{f\ reactants}$ is the entropy of formation of the reactant substances at standard conditions.)

7. $\Delta G^0_{rxn} = \sum n G^0_{f\ products} - \sum n G^0_{f\ reactants}$

This is the relationship for determining the Gibbs free energy change of a chemical reaction given the standard free energies of formation. ($\Delta G^0_{f\ products}$ is the Gibbs free energy of formation of the product substances at standard conditions. $\Delta G^0_{f\ reactants}$ is the Gibbs free energy of formation of the reactant substances at standard conditions.)

8. $\Delta G = \Delta H - T\Delta S$

This is the definition of the Gibbs free energy and is used to determine the temperature dependence of the free energy.

Heat Transfer Calculation Sample Exercises

1. *How much heat is required to heat 75.0 g of aluminum, Al, from 25.0°C to 175.0°C? The specific heat of Al is 0.900 J/g °C.*
   The correct answer is 1.01×10^4 J or 10.1 kJ.

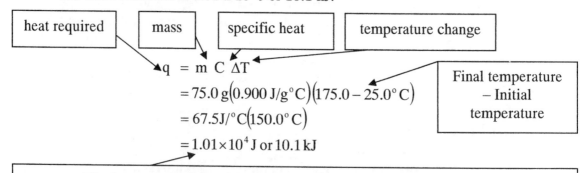

$$q = m\ C\ \Delta T$$
$$= 75.0\,g(0.900\,J/g°C)(175.0 - 25.0°C)$$
$$= 67.5J/°C(150.0°C)$$
$$= 1.01 \times 10^4\,J \text{ or } 10.1\,kJ$$

The heat required is positive indicating that the Al **absorbs** the heat.

INSIGHT: When calculating ΔT use $T_{final} - T_{initial}$. This will insure that the sign of q is correct.

2. *A 75.0 g piece of aluminum, Al, initially at a temperature of 175.0°C is dropped into a coffee cup calorimeter containing 150.0 g of water, H_2O, initially at a temperature of 15.0°C. What will the final temperature of the system be when it reaches thermal equilibrium? Assume that no heat is lost to the container. The specific heat of Al is 0.900 J/g °C and for water is 4.18 J/g°C.*
   The correct answer is 30.6°C.

INSIGHT: Heat lost = heat gained problems are characterized by the mixing of two substances at different temperatures in a common container. Be certain that you set up the problem so that the heat transfer equations are equal. You may be asked to determine the specific heat of one of the substances or the final temperature. This is an example of the latter and is the harder of the two problem types.

$$\text{heat lost by the Al} = \text{heat gained by the H}_2\text{O}$$

$$m_{Al}C_{Al}\Delta T_{Al} = m_{H_2O}C_{H_2O}\Delta T_{H_2O}$$

$$(75.0\,\text{g})(0.900\,\text{J/g}^\circ\text{C})(175.0^\circ\text{C} - T_{final}) = (150.0\,\text{g})(4.18\,\text{J/g}^\circ\text{C})(T_{final} - 15.0^\circ\text{C})$$

$$(11,812.5 - 67.5\,T)\,\text{J} = (627.0\,T - 9405.0)\,\text{J}$$

$$11,812.5 + 9405.0 = (627.0\,T + 67.5\,T)$$

$$21,217.5 = 694.5\,T$$

$$\frac{21,275}{694.5} = T$$

$$30.6^\circ\text{C} = T$$

The units of Joules, J, cancel out.

The final temperature must be between the Al's and the H_2O's temperature. Set the T_{final} so that it is between the other two temperatures.

3. **How much heat is required to convert 150.0 g of solid Al at 458°C into liquid Al at 758°C? The melting point of Al is 658°C. The specific heats for Al are, C_{solid} = 24.3 J/mol °C and C_{liquid} = 29.3 J/mol °C. The ΔH_{fusion} for Al = 10.6 kJ/mol.** The correct answer is q = 102.2 kJ.

INSIGHT: Notice that this problem involves heating a solid, melting the solid, and then heating the liquid. We must calculate each step in turn and then add them. The steps are: 1) heat required to warm the Al from 458°C to its melting point, 658°C, 2) heat required to melt the Al, and 3) heat required to heat the liquid Al from its melting point to the final temperature of 758°C. These steps are illustrated in the diagram below.

Heating Curve for Al

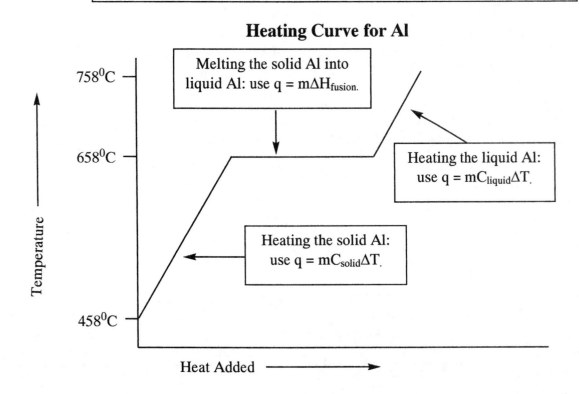

$$150.0 \text{ g}\left(\frac{1 \text{ mol Al}}{26.98 \text{ g}}\right) = 5.560 \text{ mol Al}$$

Because the ΔH and C's are in J/mol, we must convert g Al to mol.

A very common mistake is to not use the correct specific heats for each step!

Step 1) $q = mC_{solid}\Delta T = 5.560 \text{ mol}\left(24.3 \text{ J/mol } ^\circ C\right)\left(658^\circ C - 458^\circ C\right)$

$= 135 \text{ J/ } ^\circ C\left(200^\circ C\right)$

$= 2.70 \times 10^4 \text{ J} = 27.0 \text{ kJ}$

Step 2) $q = m\Delta H_{fusion}$

$= 5.560 \text{ mol}\left(10.6 \text{ kJ/mol}\right) = 58.9 \text{ kJ}$

Use the correct T range for each heating step!

Step 3) $q = mC_{liquid}\Delta T = 5.560 \text{ mol}\left(29.3 \text{ J/mol } ^\circ C\right)\left(758^\circ C - 658^\circ C\right)$

$= 163 \text{ J/}^\circ C\left(100^\circ C\right)$

$= 1.63 \times 10^4 \text{ J} = 16.3 \text{ kJ}$

Total amount of heat = Step 1 + Step 2 + Step 3 = 27.0 kJ + 58.9 kJ + 16.3 kJ = 102.2 kJ

INSIGHT: Sample exercise 3 involves only one phase change, namely converting solid Al into liquid Al. If a second phase change were included, converting solid Al into gaseous Al, the following steps would have to be included.
1) Heating the liquid Al to the boiling point using $q = mC_{liquid}\Delta T$.
2) Boiling the liquid Al using $q = m\Delta H_{vaporization}$.
3) Heating the gaseous Al using $q = mC_{gas}\Delta T$.

Change in Energy Sample Exercise

4. *If a chemical system releases 350.0 J of heat to its surroundings and has 75.0 J of work performed on it, what is the resulting change in energy of the system?*
The correct answer is -275.0 J.

INSIGHT: These are the easiest types of energy change problems. Just look for heat being released or absorbed and work being done on or by the system.

$$\Delta E = q + w$$
$$= -350.0 \text{ J} + 75.0 \text{ J}$$
$$= -275.0 \text{ J}$$

YIELD It is absolutely essential that you know the following sign conventions for heat and work.
q > 0 indicates that heat is being **absorbed** by the system
q < 0 indicates that heat is being **released** by the system
w > 0 indicates that work is being done **on** the system
w < 0 indicates that work is being done **by** the system

Work Involved in a Chemical Reaction Sample Exercise

5. *How much work is done on or by the system in the following chemical reaction at constant pressure and a temperature of $25.0^{\circ}C$?*

$$C_2H_5OH_{(\ell)} \;+\; 3\,O_{2(g)} \rightarrow 2\,CO_{2(g)} \;+\; 3\,H_2O_{(g)}$$

The correct answer is -4.957×10^3 J and the work is being done **by** the system.

INSIGHT: Look for a question that displays a chemical reaction involving a change in the number of moles of gas and asks for the amount of work performed. Be sure that you determine the change in the number of moles of gas as follows:

$$\Delta n_{gas} = \Sigma \text{moles of gas}_{products} - \Sigma \text{moles of gas}_{reactants}.$$

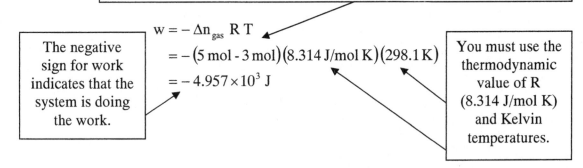

The negative sign for work indicates that the system is doing the work.

$$w = -\Delta n_{gas}\,R\,T$$
$$= -\left(5\,\text{mol} - 3\,\text{mol}\right)\left(8.314\,\text{J/mol K}\right)\left(298.1\,\text{K}\right)$$
$$= -4.957 \times 10^3 \; \text{J}$$

You must use the thermodynamic value of R (8.314 J/mol K) and Kelvin temperatures.

Relationship of Enthalpy Change, ΔH, to Energy Change, ΔE, Sample Exercise

6. *Given the following information about this chemical reaction at constant pressure and a temperature of $25.0^{\circ}C$, what are the values of ΔE, q, and w for this reaction?*

$$C_2H_5OH_{(\ell)} \;+\; 3\,O_{2(g)} \rightarrow 2\,CO_{2(g)} \;+\; 3\,H_2O_{(g)} \quad \Delta H^0_{rxn} = -1234.7 \; kJ/mol \; rxn$$

The correct answer is q = -1234.7 kJ, w = -4.957 kJ, and ΔE = -1239.3 kJ.

INSIGHT: Exercise 4 tells us that the value of w for this reaction is -4.957×10^3 J or -4.957 kJ. The definition of enthalpy change, $\Delta H = \Delta E + P\Delta V = q_P$ gives us a method to determine q. **The $\Delta H^0_{rxn} = q$ at constant pressure.** Thus q = -1234.7 kJ.

$$\Delta E = q + w$$
$$= \left(-1234.7\,\text{kJ}\right) + \left(-4.957\,\text{kJ}\right)$$
$$= -1239.7 \; \text{kJ}$$

YIELD Notice that ΔH and ΔE are almost the same value. They differ only by the amount of work that the system does, -4.957 kJ. This agrees with the definition of enthalpy change, $\Delta H = \Delta E + P\Delta V = \Delta E + \Delta n_{gas}RT$ at constant temperature and pressure.

Calculation of Enthalpy Change for a Reaction Sample Exercise

7. *What is the enthalpy change for this reaction at standard conditions?*

$$C_2H_5OH_{(l)} + 3\,O_{2(g)} \rightarrow 2\,CO_{2(g)} + 3\,H_2O_{(g)}$$

The correct answer is $\Delta H^0_{rxn} = -1234.7$ kJ/mol rxn.

The superscript 0's indicate that these values were measured at standard thermodynamic conditions (1.00 atm of pressure and 273.15 K). The n's are the stoichiometric coefficients from the balanced reaction.	Notice that we must sum the products' ΔH values and the reactants' ΔH values then subtract them. The subscript f's indicate that the ΔH values are for the formation of the substances from their elements.	The ΔH^0_f values are tabulated in an appendix at the back of your textbook. Elements, like O_2, always have a $\Delta H^0_f = 0.0$ kJ/mol.

$$\Delta H^0_{rxn} = \sum n\Delta H^0_{f\ products} - \sum n\Delta H^0_{f\ reactants}$$

$$= \Big[\underbrace{2(-393.5\ \text{kJ/mol})}_{2\ CO_2} + \underbrace{3(-241.8\ \text{kJ/mol})}_{3\ H_2O}\Big] - \Big[\underbrace{1(-277.7\ \text{kJ/mol})}_{1\ C_2H_5OH} + \underbrace{3(0.0\ \text{kJ/mol})}_{3\ O_2}\Big]$$

$$= \big[-787.0\ \text{kJ/mol} - 725.4\ \text{kJ/mol}\big] - \big[-277.7\ \text{kJ/mol}\big]$$

$$= -1512.4\ \text{kJ/mol} + 277.7\ \text{kJ/mol}$$

$$= -1234.7\ \text{kJ/mol}$$

The negative value for ΔH^0_{rxn} indicates that this is an **exothermic** reaction.	Be very careful with the signs of the ΔH^0_f values and how they are added and subtracted.	The brackets are to help you understand how the ΔH^0_f values and stoichiometric coefficients are determined.

Calculation of Entropy Change for a Reaction Sample Exercise

8. *What is the entropy change for this reaction at standard conditions?*

$$C_2H_5OH_{(l)} + 3\,O_{2(g)} \rightarrow 2\,CO_{2(g)} + 3\,H_2O_{(g)}$$

The correct answer is $\Delta S^0_{rxn} = 217.3$ J/mol

The n's, superscript 0's, and subscript f's have the same meaning in this equation as in exercise 6. The ΔS^0_f values are also tabulated in an appendix in your text.

$$\Delta S^0_{rxn} = \sum n\Delta S^0_{f\ products} - \sum n\Delta S^0_{f\ reactants}$$

$$= \left[\underbrace{2(213.6\ \text{J/mol K})}_{2\ CO_2} + \underbrace{3(188.7\ \text{J/mol K})}_{3\ H_2O}\right] - \left[\underbrace{1(161.0\ \text{J/mol K})}_{1\ C_2H_5OH} + \underbrace{3(205.0\ \text{J/mol K})}_{3\ O_2}\right]$$

$$= [427.2\ \text{J/mol K} + 566.1\ \text{J/mol K}] - [161.0\ \text{J/mol K} + 615.0\ \text{J/mol K}]$$

$$= 993.3\ \text{J/mol K} + 776.0\ \text{J/mol K}$$

$$= 217.3\ \text{J/mol K}$$

The positive value for ΔS^0_{rxn} indicates that this chemical system is more disordered after the reaction has occurred.

Elements can have nonzero values of ΔS^0_f.

Calculation of Gibbs Free Energy Change for a Reaction Sample Exercise

9. *What is the Gibbs Free Energy change for this reaction at standard conditions?*

$$C_2H_5OH_{(l)} + 3\,O_{2(g)} \rightarrow 2\,CO_{2(g)} + 3\,H_2O_{(g)}$$

The correct answer is $\Delta G^0_{rxn} = -1299.7$ kJ/mol.

The ΔG^0_f values are also tabulated in an appendix in your textbook.

$$\Delta G^0_{rxn} = \sum n\Delta G^0_{f\ products} - \sum n\Delta G^0_{f\ reactants}$$

$$= \left[\underbrace{2(-394.4\ \text{kJ/mol})}_{2\ CO_2} + \underbrace{3(-228.6\ \text{kJ/mol})}_{3\ H_2O}\right] - \left[\underbrace{1(-174.9\ \text{kJ/mol})}_{1\ C_2H_5OH} + \underbrace{3(0.0\ \text{kJ/mol})}_{3\ O_2}\right]$$

$$= [-788.8\ \text{kJ/mol} - 685.8\ \text{kJ/mol}] - [-174.9\ \text{J/mol} + 0.0\ \text{kJ/mol}]$$

$$= -1474.6\ \text{kJ/mol} + 174.9\ \text{kJ/mol}$$

$$= -1299.7\ \text{kJ/mol}$$

The negative sign for ΔG^0_{rxn} indicates that this reaction is spontaneous.

Elements have zero values of ΔG^0_f.

It is very important that you know the following sign conventions for ΔH, ΔS, and ΔG.

$\Delta H > 0$ indicates that the process is **endothermic**.
$\Delta H < 0$ indicates that the process is **exothermic**.
$\Delta S > 0$ indicates that the process is **less ordered**.
$\Delta S < 0$ indicates that the process is **more ordered**.
$\Delta G > 0$ indicates that the process is **nonspontaneous**.
$\Delta G < 0$ indicates that the process is **spontaneous**.

Temperature Dependence of Spontaneity Sample Exercise

10. Can this reaction become nonspontaneous if the temperature is changed?

$$C_2H_5OH_{(l)} \; + \; 3 \, O_{2(g)} \rightarrow 2 \, CO_{2(g)} \; + \; 3 \, H_2O_{(g)}$$

The correct answer is no.

INSIGHT:

From exercises 6 and 7 we see that for this reaction $\Delta H^0_{rxn} < 0$ and $\Delta S^0_{rxn} > 0$. The definition of ΔG^0_{rxn} is $\Delta G^0_{rxn} = \Delta H^0_{rxn} - T\Delta S^0_{rxn}$. In this case, $\Delta G^0_{rxn} =$ (negative quantity) $- T$ (positive value) which must always give $\Delta G^0_{rxn} < 0$. (Remember, $\Delta G^0_{rxn} < 0$ indicates a spontaneous reaction.)

The following conclusions can be drawn regarding the temperature dependence of ΔG^0_{rxn}.

If $\Delta H^0_{rxn} < 0$ and $\Delta S^0_{rxn} > 0$, then $\underline{\Delta G^0_{rxn} < 0 \text{ at all temperatures}}$.
If $\Delta H^0_{rxn} > 0$ and $\Delta S^0_{rxn} < 0$, then $\underline{\Delta G^0_{rxn} > 0 \text{ at all temperatures}}$.
If $\Delta H^0_{rxn} < 0$ and $\Delta S^0_{rxn} < 0$, then $\underline{\Delta G^0_{rxn} < 0 \text{ at low temperatures}}$.
If $\Delta H^0_{rxn} > 0$ and $\Delta S^0_{rxn} > 0$, then $\underline{\Delta G^0_{rxn} < 0 \text{ at high temperatures}}$.

Module 16
Chemical Kinetics

Introduction
Chemical kinetics describes how quickly chemical reactions occur. There are several factors that can be humanly controlled to change the rate of a reaction including temperature and the concentrations of the reactants. This module will describe the relationship of the rates of one reactant to the rates of other reactants and products, how to determine the order of a reactant from experimental data, integrated rate laws for first and second order reactions, and the effect of temperature on the rate of a reaction using the Arrhenius equation.

Module 16 Key Equations & Concepts

1. $\text{rate} \propto \dfrac{-\Delta[A]}{\Delta t} = \dfrac{-\Delta[B]}{b\Delta t} = \dfrac{+\Delta[C]}{c\Delta t}$ **for the reaction** $A + bB \rightarrow cC$

    This is the definition of the rate of a reaction based on the concentrations of the reactants or products. (The symbol [A] represents the molar concentration of substance A and similarly for [B] and [C].) Notice that the concentrations of A and B (reactants) decrease with time, t, and that of C (a product) increases with time.

2. $[A] = [A_0]e^{-kt}$

    This is the integrated rate law for chemical reactions that obey first order kinetics. It is used to determine the concentration of a reactant a certain amount of time after a reaction has started or the amount of time required for the concentration of a reactant to reach a specified amount. [A] is the concentration of A after time has passed, $[A_0]$ is initial concentration of A, k is the rate constant, and t is the amount of time.

3. $k\, t_{1/2} = 0.693$

    The half-life relationship for first order reactions is a method to determine the half-life of a first order reaction given the rate constant or vice versa. The half-life is the length of time for the concentration to reach one-half the initial amount.

4. $\dfrac{1}{[A]} - \dfrac{1}{[A_0]} = kt$

    This is the integrated rate law for chemical reactions that obey second order kinetics. It is used to determine the concentration of a reactant a certain amount of time after a reaction has started or the amount of time required for the concentration of a reactant to reach a specified amount.

5. $k\, t_{1/2} = \dfrac{1}{[A_0]}$

    The half-life relationship for second order reactions is a method to determine the half-life of a second order reaction given the rate constant or vice versa.

6. $$\ln \frac{k_2}{k_1} = \frac{E_a}{R}\left(\frac{1}{T_1} - \frac{1}{T_2}\right)$$

The Arrhenius equation describes how the rate of a reaction changes when the reaction temperature is increased or decreased. E_a is the activation energy of the reaction; k_2 and k_1 are the rate constants at temperatures T_1 and T_2; R is the universal gas constant.

Rates of a Reaction Based on the Concentrations of Products and Reactants Sample Exercise

1. *How are the rates of the disappearance of O_2 and the appearance of H_2O related to the rate of disappearance of H_2 in this reaction?*

$$2\,H_{2(g)} + O_{2(g)} \rightarrow 2\,H_2O_{(g)}$$

The correct answer is $\dfrac{-\Delta[H_2]}{\Delta t} = \dfrac{-2\Delta[O_2]}{\Delta t} = \dfrac{+\Delta[H_2O]}{\Delta t}$.

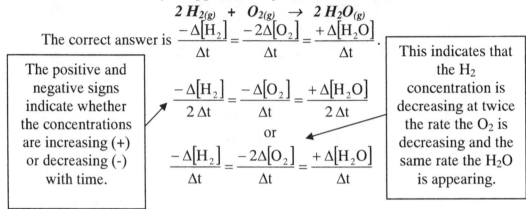

The positive and negative signs indicate whether the concentrations are increasing (+) or decreasing (-) with time.

$$\frac{-\Delta[H_2]}{2\,\Delta t} = \frac{-\Delta[O_2]}{\Delta t} = \frac{+\Delta[H_2O]}{2\,\Delta t}$$

or

$$\frac{-\Delta[H_2]}{\Delta t} = \frac{-2\Delta[O_2]}{\Delta t} = \frac{+\Delta[H_2O]}{\Delta t}$$

This indicates that the H_2 concentration is decreasing at twice the rate the O_2 is decreasing and the same rate the H_2O is appearing.

Determination of the Order of a Reaction from Experimental Data Sample Exercise

2. *For this chemical reaction:*

$$(C_2H_5)_3N + C_2H_5Br \rightarrow (C_2H_5)_4NBr$$

the following experimental data were obtained.

Experiment	[(C₂H₅)₃N] (*M*)	[C₂H₅Br] (*M*)	Relative rate (*M/min*)
1	0.10	0.10	3.0
2	0.20	0.10	6.0
3	0.10	0.30	9.0

What is the rate equation for this reaction and the value of the rate constant, k?
The correct answer is rate = k $[(C_2H_5)_3N]^1$ $[C_2H_5Br]^1$ and k = 3.0 x 10^2 M^{-1} min^{-1}.

INSIGHT: Problems of this type will present a set of data in which the concentration of one of the reactants changes and the other reactants' concentrations remain constant. For example, compare experiments 1 and 2. Notice that the [(C₂H₅)₃N] doubles, 0.10 *M* to 0.20 *M*, and the [C₂H₅Br] remains constant at 0.10 *M*. Thus the concentration effects on the rate have been isolated to the [(C₂H₅)₃N]. Now, look at the relative rates for experiments 1 and 2 which changes from 3.0 to 6.0 *M/*min, i.e. it also doubles. This indicates that the reaction is 1st order with respect to [(C₂H₅)₃N].

INSIGHT:

Now compare experiments 1 and 3 where the $[(C_2H_5)_3N]$ remains constant at 0.10 M and the $[C_2H_5Br]$ triples, from 0.10 M to 0.30 M. Notice that the rate also triples from 3.0 to 9.0 M/min. Thus we can conclude that this reaction is 1st order with respect to $[C_2H_5Br]$. That is the information required to write the form of the rate law for this reaction. Thus we can conclude that this reaction is 1st order with respect to $[C_2H_5Br]$. That is the information required to write the form of the rate law for this reaction.

rate = k $[(C_2H_5)_3N]^1$ $[C_2H_5Br]^1$

We say that this reaction is <u>1st order with respect to $[(C_2H_5)_3N]$</u>, <u>1st order with respect to $[C_2H_5Br]$</u> and <u>2nd order overall</u>.
(The overall order is the sum of the individual orders.)

YIELD

A very common mistake is to assume that the order of a reaction is determined by the stoichiometric coefficients of the balanced chemical reaction. This is not correct. **The only method to determine the order of a reaction is analysis of experimental data just as is done in this problem.**

The values of the rate and the concentrations of both $(C_2H_5)_3N$ and C_2H_5Br come from experiment 3.

The value of the rate constant, k, can be determined from the data from experiments 1, 2, or 3. As an example let us choose experiment 3's data.

$$\text{rate} = k[(C_2H_5)_3N]^1[(C_2H_5Br)]^1 \text{ thus}$$

$$k = \frac{\text{rate}}{[(C_2H_5)_3N]^1[(C_2H_5Br)]^1}$$

$$k = \frac{9.0\,M/\min}{(0.10\,M)(0.30\,M)} = 3.0\times10^2\,\frac{1}{M\,\min}$$

Units for k of $1/(M \times time)$ are always correct for 2nd order reactions.

First Order Integrated Rate Law Sample Exercises

3. *The following reaction is first order with respect to $[NH_2NO_2]$ and the value of the rate constant, k, is 9.3×10^{-5} s^{-1}. If the initial $[NH_2NO_2] = 2.0$ M, what will the $[NH_2NO_2]$ be 30.0 minutes after the reaction has started?*

$$NH_2NO_{2(aq)} \rightarrow N_2O_{(g)} + H_2O_{(\ell)}$$

The correct answer is 1.7 M.

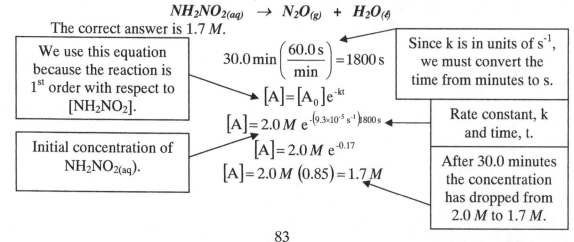

We use this equation because the reaction is 1st order with respect to $[NH_2NO_2]$.

Initial concentration of $NH_2NO_{2(aq)}$.

$$30.0\,\min\left(\frac{60.0\,s}{\min}\right) = 1800\,s$$

$$[A] = [A_0]e^{-kt}$$

$$[A] = 2.0\,M\,e^{-(9.3\times10^{-5}\,s^{-1})1800\,s}$$

$$[A] = 2.0\,M\,e^{-0.17}$$

$$[A] = 2.0\,M\,(0.85) = 1.7\,M$$

Since k is in units of s^{-1}, we must convert the time from minutes to s.

Rate constant, k and time, t.

After 30.0 minutes the concentration has dropped from 2.0 M to 1.7 M.

4. *The following reaction is first order with respect to [NH₂NO₂] and the value of the rate constant, k, is 9.3 x 10⁻⁵ s⁻¹. If the initial [NH₂NO₂] = 2.0 M, how long will it be before the [NH₂NO₂] = 1.5 M?*

$$NH_2NO_{2(aq)} \rightarrow N_2O_{(g)} + H_2O_{(\ell)}$$

The correct answer is 3.1×10^3 s or 52 min.

INSIGHT:	This problem is a slight variation of exercise 3. All that is required is a little algebra to solve the integrated rate law for t instead of A.

YIELD

For reactions that obey simple first order kinetics, i.e. rate = k [A]¹, the following important points must be remembered:

1) The units of the rate constant, k, will always be 1/time. For example they might be 1/s or 1/min or 1/yr. These units can also be written as s^{-1}, min^{-1}, or yr^{-1}.

2) A very common mistake is to assume that for first order reactions the concentration decreases linearly, in other words as a simple ratio. **First order reaction concentrations decrease exponentially not linearly!**

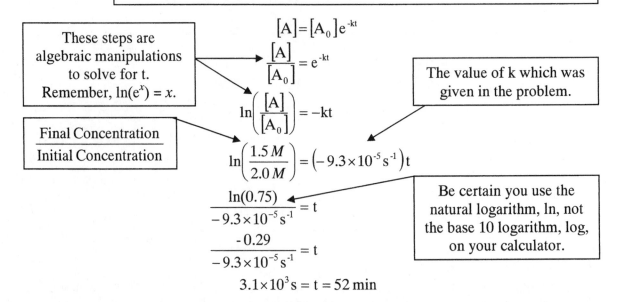

These steps are algebraic manipulations to solve for t. Remember, $\ln(e^x) = x$.

Final Concentration
Initial Concentration

The value of k which was given in the problem.

$$[A] = [A_0] e^{-kt}$$

$$\frac{[A]}{[A_0]} = e^{-kt}$$

$$\ln\left(\frac{[A]}{[A_0]}\right) = -kt$$

$$\ln\left(\frac{1.5\,M}{2.0\,M}\right) = \left(-9.3 \times 10^{-5}\,s^{-1}\right)t$$

$$\frac{\ln(0.75)}{-9.3 \times 10^{-5}\,s^{-1}} = t$$

$$\frac{-0.29}{-9.3 \times 10^{-5}\,s^{-1}} = t$$

$$3.1 \times 10^3\,s = t = 52\,min$$

Be certain you use the natural logarithm, ln, not the base 10 logarithm, log, on your calculator.

5. *The following reaction is first order with respect to [NH₂NO₂] and the value of the rate constant, k, is 9.3 x 10⁻⁵ s⁻¹. What is the half-life of this reaction?*

$$NH_2NO_{2(aq)} \rightarrow N_2O_{(g)} + H_2O_{(\ell)}$$

The correct answer is 7.5×10^3 s or 1.2×10^2 min.

$$kt_{1/2} = 0.693$$

$$t_{1/2} = \frac{0.693}{k}$$

$$t_{1/2} = \frac{0.693}{9.3 \times 10^{-5}\,s^{-1}}$$

$$t_{1/2} = 7.5 \times 10^3\,s = 1.2 \times 10^2\,min$$

We use this equation because the reaction is 1st order with respect to [NH₂NO₂].

Second Order Integrated Rate Law Sample Exercises

6. *The following reaction at 400.0 K is second order with respect to [CF₃] and the value of the rate constant, k, is 2.51 x 10¹⁰ M⁻¹s⁻¹. If the initial [CF₃] = 2.0 M, what will the [CF₃] be 4.25 x 10⁻¹⁰ seconds after the reaction has started?*

$$2\ CF_{3(g)}\ \rightarrow\ C_2F_{6(g)}$$

The correct answer is 8.96×10^{-2} M.

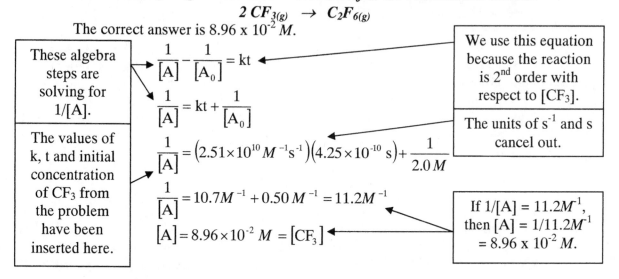

These algebra steps are solving for 1/[A].	$\dfrac{1}{[A]} - \dfrac{1}{[A_0]} = kt$	We use this equation because the reaction is 2ⁿᵈ order with respect to [CF₃].
	$\dfrac{1}{[A]} = kt + \dfrac{1}{[A_0]}$	The units of s⁻¹ and s cancel out.
The values of k, t and initial concentration of CF₃ from the problem have been inserted here.	$\dfrac{1}{[A]} = \left(2.51 \times 10^{10}\,M^{-1}s^{-1}\right)\left(4.25 \times 10^{-10}\,s\right) + \dfrac{1}{2.0\,M}$	
	$\dfrac{1}{[A]} = 10.7 M^{-1} + 0.50\,M^{-1} = 11.2 M^{-1}$	If $1/[A] = 11.2 M^{-1}$, then $[A] = 1/11.2 M^{-1}$ = 8.96 x 10⁻² M.
	$[A] = 8.96 \times 10^{-2}\,M = [CF_3]$	

7. *The following reaction at 400.0 K is second order with respect to [CF₃] and the value of the rate constant, k, is 2.51 x 10¹⁰ M⁻¹s⁻¹. If the initial [CF₃] = 2.0 M, how long will it be before the [CF₃] = 1.5 M?*

$$2\ CF_{3(g)}\ \rightarrow\ C_2F_{6(g)}$$

The correct answer is 6.8×10^{-12} s.

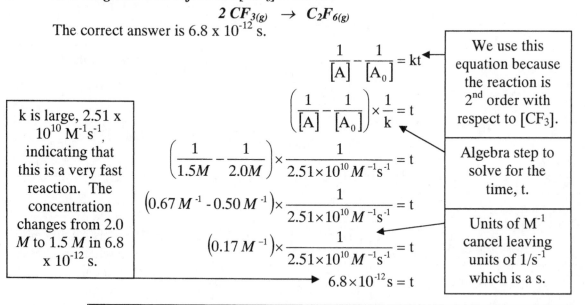

k is large, 2.51 x 10¹⁰ M⁻¹s⁻¹, indicating that this is a very fast reaction. The concentration changes from 2.0 M to 1.5 M in 6.8 x 10⁻¹² s.	$\dfrac{1}{[A]} - \dfrac{1}{[A_0]} = kt$	We use this equation because the reaction is 2ⁿᵈ order with respect to [CF₃].
	$\left(\dfrac{1}{[A]} - \dfrac{1}{[A_0]}\right) \times \dfrac{1}{k} = t$	Algebra step to solve for the time, t.
	$\left(\dfrac{1}{1.5M} - \dfrac{1}{2.0M}\right) \times \dfrac{1}{2.51 \times 10^{10}\,M^{-1}s^{-1}} = t$	Units of M⁻¹ cancel leaving units of 1/s⁻¹ which is a s.
	$\left(0.67\,M^{-1} - 0.50\,M^{-1}\right) \times \dfrac{1}{2.51 \times 10^{10}\,M^{-1}s^{-1}} = t$	
	$\left(0.17\,M^{-1}\right) \times \dfrac{1}{2.51 \times 10^{10}\,M^{-1}s^{-1}} = t$	
	$6.8 \times 10^{-12}\,s = t$	

YIELD

For reactions that obey simple second order kinetics, i.e. rate = k [A]², the units of k will always be $\dfrac{1}{(concentration)(time)}$. For example, k could be in any of these units: $\dfrac{1}{M\ s}$ or $\dfrac{1}{M\ min}$ or $\dfrac{1}{M\ yr}$ which could also be written as $M^{-1}\,s^{-1}$, $M^{-1}\,min^{-1}$, $M^{-1}\,yr^{-1}$.

8. **The following reaction at 400 K is second order with respect to [CF₃] and the value of the rate constant, k, is 2.51 x 10¹⁰ M⁻¹s⁻¹. If the initial [CF₃] = 2.0 M, what is the half-life of the reaction?**

$$2\ CF_{3(g)} \rightarrow C_2F_{6(g)}$$

The correct answer is 6.8×10^{-12} s.

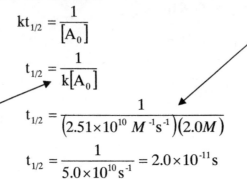

Unlike first order reactions, the half-life for second order reactions changes with the initial concentration of the reactant.

The unit of M^{-1} cancels with M leaving units of $1/s^{-1}$ which is a s.

$$kt_{1/2} = \frac{1}{[A_0]}$$

$$t_{1/2} = \frac{1}{k[A_0]}$$

$$t_{1/2} = \frac{1}{(2.51 \times 10^{10}\ M^{-1}s^{-1})(2.0M)}$$

$$t_{1/2} = \frac{1}{5.0 \times 10^{10}\ s^{-1}} = 2.0 \times 10^{-11}s$$

Arrhenius Equation Sample Exercises

9. **A reaction has an activation energy of 52.0 kJ/mol and a rate constant, k, of 7.50×10^2 s⁻¹ at 300.0 K. What is the rate constant for this reaction at 350.0 K?**

The correct answer is 1.42×10^4 s⁻¹.

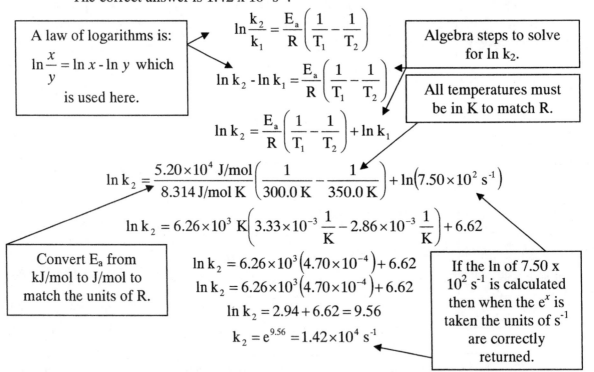

A law of logarithms is:

$$\ln \frac{x}{y} = \ln x - \ln y \text{ which}$$

is used here.

$$\ln \frac{k_2}{k_1} = \frac{E_a}{R}\left(\frac{1}{T_1} - \frac{1}{T_2}\right)$$

Algebra steps to solve for ln k₂.

$$\ln k_2 - \ln k_1 = \frac{E_a}{R}\left(\frac{1}{T_1} - \frac{1}{T_2}\right)$$

All temperatures must be in K to match R.

$$\ln k_2 = \frac{E_a}{R}\left(\frac{1}{T_1} - \frac{1}{T_2}\right) + \ln k_1$$

$$\ln k_2 = \frac{5.20 \times 10^4\ \text{J/mol}}{8.314\ \text{J/mol K}}\left(\frac{1}{300.0\ K} - \frac{1}{350.0\ K}\right) + \ln(7.50 \times 10^2\ s^{-1})$$

$$\ln k_2 = 6.26 \times 10^3\ K\left(3.33 \times 10^{-3}\frac{1}{K} - 2.86 \times 10^{-3}\frac{1}{K}\right) + 6.62$$

Convert E_a from kJ/mol to J/mol to match the units of R.

$$\ln k_2 = 6.26 \times 10^3(4.70 \times 10^{-4}) + 6.62$$

$$\ln k_2 = 6.26 \times 10^3(4.70 \times 10^{-4}) + 6.62$$

$$\ln k_2 = 2.94 + 6.62 = 9.56$$

$$k_2 = e^{9.56} = 1.42 \times 10^4\ s^{-1}$$

If the ln of 7.50 x 10^2 s⁻¹ is calculated then when the e^x is taken the units of s⁻¹ are correctly returned.

INSIGHT: Kinetics problems that deal with changing rates, or rate constants (k), and temperature changes require use of the Arrhenius equation.

10. What is the activation energy of a reaction that has a rate constant of 2.50 x 10^2 kJ/mol at 325K and a rate constant of 5.00 x 10^2 kJ/mol at 375 K?
   The correct answer is 26 kJ/mol.

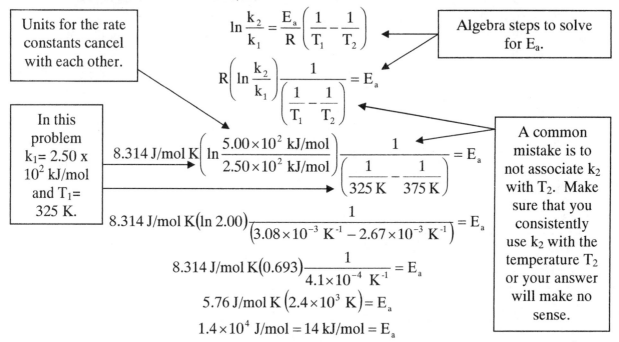

Units for the rate constants cancel with each other.

$$\ln \frac{k_2}{k_1} = \frac{E_a}{R}\left(\frac{1}{T_1} - \frac{1}{T_2}\right)$$

Algebra steps to solve for E_a.

$$R\left(\ln \frac{k_2}{k_1}\right)\frac{1}{\left(\dfrac{1}{T_1} - \dfrac{1}{T_2}\right)} = E_a$$

In this problem k_1= 2.50 x 10^2 kJ/mol and T_1= 325 K.

$$8.314\ \text{J/mol K}\left(\ln \frac{5.00\times 10^2\ \text{kJ/mol}}{2.50\times 10^2\ \text{kJ/mol}}\right)\frac{1}{\left(\dfrac{1}{325\,\text{K}} - \dfrac{1}{375\,\text{K}}\right)} = E_a$$

A common mistake is to not associate k_2 with T_2. Make sure that you consistently use k_2 with the temperature T_2 or your answer will make no sense.

$$8.314\ \text{J/mol K}(\ln 2.00)\frac{1}{\left(3.08\times 10^{-3}\ \text{K}^{-1} - 2.67\times 10^{-3}\ \text{K}^{-1}\right)} = E_a$$

$$8.314\ \text{J/mol K}(0.693)\frac{1}{4.1\times 10^{-4}\ \text{K}^{-1}} = E_a$$

$$5.76\ \text{J/mol K}\left(2.4\times 10^3\ \text{K}\right) = E_a$$

$$1.4\times 10^4\ \text{J/mol} = 14\ \text{kJ/mol} = E_a$$

Module 17
Gas Phase Equilibria

Introduction

This module is a description of the basic calculations required for the simplest equilibrium problems, gas phase equilibria. In this module we will determine the value of K_c, the equilibrium constant, and use it to predict if a reaction is product or reactant favored, and to calculate the concentrations of species in a reaction; see how K_P, the equilibrium constant in terms of the partial pressures of the gases, can be calculated and how it is related to K_c; use Le Chatelier's Principle and the reaction quotient, Q, to predict the effects of temperature, pressure, and concentration changes on an equilibrium; examine the relationship of ΔG and K; and calculate the value of K at different temperatures. Many of the ideas introduced here will be used again in Module 18 with slight variations.

Module 17 Key Equations & Concepts

1. $K_c = \dfrac{[C]^c[D]^d}{[A]^a[B]^b}$ for the reaction $aA + bB \rightleftarrows cC + dD$

    The equilibrium constant, K_c, is used to determine whether the reaction is product or reactant favored (i.e. yields a preponderance of reactants or products) and to determine the concentrations of the reactants and products at equilibrium. Concentrations used in K_c must be equilibrium concentrations. The symbols a, b, c, and d are the stoichiometric coefficients for the reaction.

2. $K_P = \dfrac{(P_C)^c(P_D)^d}{(P_A)^a(P_B)^b}$ for the reaction $aA + bB \rightleftarrows cC + dD$

    where all species are in the gas phase.

    The equilibrium constant in terms of the partial pressures of the gases, K_P, serves the same purpose as K_c except that the partial pressures of the gases are used instead of the equilibrium concentrations. K_P is used when it is easiest to measure the pressures of the gases instead of the equilibrium concentrations.

3. $Q = \dfrac{[C]^c[D]^d}{[A]^a[B]^b}$ for the reaction $aA + bB \rightleftarrows cC + dD$

    The reaction quotient, Q, has the same form as K_c but the concentrations are all nonequilibrium. Q is used to determine how the position of equilibrium must shift for a nonequilibrium system to attain equilibrium.

4. $K_P = K_c(RT)^{\Delta n}$ where

    $\Delta n = \sum$ **moles of gaseous products - $\sum$ moles of gaseous reactants**

    This relationship defines how the partial pressure equilibrium constant, K_P, and the equilibrium constant, K_c, are related.

5. $\Delta G^0_{rxn} = -RT \ln K$

    K, the thermodynamic equilibrium constant, is related to the standard Gibbs Free Energy change using this relationship.

---

6. $$\ln\left(\frac{K_{T_2}}{K_{T_1}}\right) = \frac{\Delta H^0}{R}\left(\frac{1}{T_1} - \frac{1}{T_2}\right)$$

The van't Hoff equation describes how to determine the values of equilibrium constants at different temperatures.

Use of the Equilibrium Constant, K_c, Sample Exercises

1. *For the following reaction at 298 K, the equilibrium concentrations are [H₂] = 1.50 M, [I₂] = 2.00 M, and [HI] = 3.46 M. What is the value of the equilibrium constant, K_c, for this reaction at 298 K?*

$$H_{2(g)} + I_{2(g)} \rightleftarrows 2HI_{(g)}$$

The correct answer is 4.00.

Units are not used in equilibrium constants. We are interested in K_c's size.	For this reaction $K_c = \dfrac{[HI]^2}{[H_2][I_2]}$ thus $K_c = \dfrac{[3.46]^2}{[1.50][2.00]} = \dfrac{12.0}{3.00} = 4.00$	A very common mistake is to forget to properly include the stoichiometric coefficients as exponents.

YIELD

The size of an equilibrium constant indicates if the reaction yields a preponderance of products, reactants, or neither.
1) If K_c >10 to 20, the reaction is a **product favored reaction**.
2) If K_c < 1, the reaction is a **reactant favored reaction**.
3) If 1 < K_c < 10 to 20, the reaction yields a **mixture of reactants and products**.

INSIGHT: Fundamentally, K_c is a ratio of the product concentrations divided by the reactant concentrations. This is why the larger the value of K_c the more product favored the reaction is.

INSIGHT: K_c is actually defined using a thermodynamic quantity called activity. The activities of gases are the same as their concentrations. In heterogeneous equilibria (those involving gases, liquids, and solids) the activities of the pure solids and liquids are 1 and can be neglected. Thus the K_c for heterogeneous equilibria will only require the gas's concentrations and not any solids or liquids involved in the equilibrium.

2. *For the following reaction at 298 K, the equilibrium constant is 4.00. If the reaction vessel initially has the following reactant concentrations [H₂] = 6.00 M and [I₂] = 4.00 M. What are the equilibrium concentrations of all species in this reaction?*

$$H_{2(g)} + I_{2(g)} \rightleftarrows 2HI_{(g)}$$

The correct answer is [H₂] = 3.6 *M*, [I₂] = 1.6 *M*, and [HI] = 4.8 *M*.

Problems that give you the K_c and the starting concentrations of the reactants must be solved in this fashion. Set up this table to insure success.

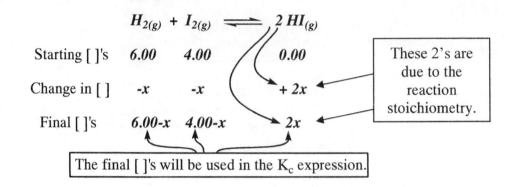

$$K_c = \frac{[HI]^2}{[H_2][I_2]} = 4.00$$

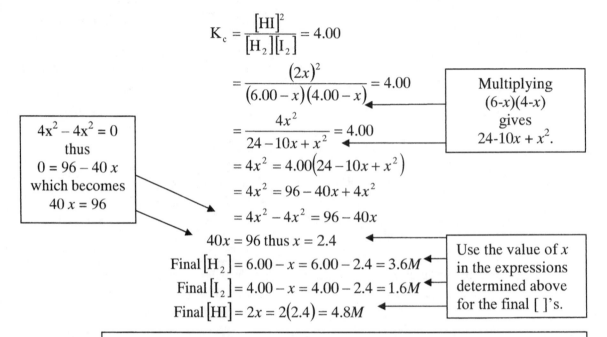

$$= \frac{(2x)^2}{(6.00-x)(4.00-x)} = 4.00$$

Multiplying $(6-x)(4-x)$ gives $24-10x+x^2$.

$$= \frac{4x^2}{24-10x+x^2} = 4.00$$

$4x^2 - 4x^2 = 0$
thus
$0 = 96 - 40\,x$
which becomes
$40\,x = 96$

$$= 4x^2 = 4.00(24-10x+x^2)$$

$$= 4x^2 = 96 - 40x + 4x^2$$

$$= 4x^2 - 4x^2 = 96 - 40x$$

$40x = 96$ thus $x = 2.4$

Use the value of x in the expressions determined above for the final [ ]'s.

Final $[H_2] = 6.00 - x = 6.00 - 2.4 = 3.6M$
Final $[I_2] = 4.00 - x = 4.00 - 2.4 = 1.6M$
Final $[HI] = 2x = 2(2.4) = 4.8M$

INSIGHT: In this particular problem the x^2 terms cancel out simplifying the problem. If this had not happened, the problem would have been solved using the quadratic equation.

Use of the Equilibrium Constant, K_P, Sample Exercises

3. *For the following reaction at 298 K, the equilibrium partial pressures are*
   $P_{NO_2} = 0.500$ *atm and* $P_{N_2O_4} = 0.0698$ *atm. What is the value of K_P for this reaction?*

$$2\,NO_{2(g)} \rightleftharpoons N_2O_{4(g)}$$

The correct answer is $K_P = 0.279$.

$$K_P = \frac{[N_2O_4]}{[NO_2]^2} = \frac{0.0698}{(0.500)^2} = \frac{0.0698}{0.250} = 0.279$$

The stoichiometric coefficients are also used in K_P calculations.

4. *For the following reaction at 298 K, K_P has the value of 0.279, what is the value of K_c for this reaction at 298 K?*

$$2\,NO_{2(g)} \rightleftarrows N_2O_{4(g)}$$

The correct answer is $K_c = 6.84$.

$$K_P = K_c(RT)^{\Delta n} \text{ thus } K_c = \frac{K_P}{(RT)^{\Delta n}}$$

$\Delta n = \sum \text{moles of gaseous products} - \sum \text{moles of gaseous reactants}$

$\Delta n = \text{moles of } N_2O_4 - \text{moles of } NO_2 = 1 - 2 = -1$

$$\frac{1}{(0.0821 \times 298)^{-1}} =$$
$$\frac{1}{(24.5)^{-1}} = 24.5$$

Use the gas law value of R (0.0821 L atm/mol K).

$$K_c = \frac{0.279}{(0.0821 \times 298)^{-1}} = 0.279 \times 24.5 = 6.84$$

Effects of Temperature, Pressure, and Concentration on the Position of Equilibrium
Sample Exercise

5. *What would be the effect of each of these changes on the position of equilibrium of this reaction at 298 K?*

$$2\,NO_{2(g)} \rightleftarrows N_2O_{4(g)} \qquad \Delta H^0_{rxn} = -57.2\ kJ/mol$$

   a) *Increasing the temperature of the reaction*
   b) *Removing some NO_2 from the reaction vessel.*
   c) *Adding some N_2O_4 to the reaction vessel.*
   d) *Increasing the pressure in the reaction vessel by adding an inert gas.*
   e) *Decreasing by half the size of the reaction vessel.*
   f) *Introducing a catalyst into the reaction vessel.*
The correct answers are the position of equilibrium will shift to the: a) left b) left c) left d) no effect e) right f) no effect.

INSIGHT: If the position of equilibrium shifts to the left the reactant concentrations increase and the product concentrations decrease. If the position of equilibrium shifts to the right the reactant concentrations decrease and the product concentrations increase. All of these changes are illustrations of Le Chatelier's principle: When a system at equilibrium is stressed, it will respond to relieve that stress.

   a) *For exothermic reactions: increasing the temperature shifts the position of equilibrium to the left, decreasing the temperature shifts the position of equilibrium the right. Endothermic reactions behave oppositely.* In this exercise the negative ΔH^0_{rxn} tells us that the reaction is exothermic, thus increasing the temperature shifts the position of equilibrium to the left.

YIELD

b) *If a reactant's concentration is decreased below the equilibrium concentration, the position of equilibrium will change to restore concentrations that correspond to those predicted by the equilibrium constant.* In this exercise, removing some NO_2 from the reaction vessel decreases the $[NO_2]$. The reaction equilibrium responds to this stress by increasing the $[NO_2]$ and decreasing the $[N_2O_4]$, an equilibrium position shift to the left or reactant side. Adding NO_2 would cause the position of equilibrium to shift to the right or product side.

c) *If a product's concentration is increased above the equilibrium concentration, the equilibrium position will shift to restore concentrations of products and reactants that correspond to those predicted by the equilibrium constant.* Adding some N_2O_4 to the reaction vessel increases the $[N_2O_4]$ above the equilibrium concentration. The reaction equilibrium responds to this stress by decreasing the $[N_2O_4]$ and increasing the $[NO_2]$, an equilibrium position shift to the left or reactant side. Removing N_2O_4 would shift the position of equilibrium to the right or product side.

d) *Adding an inert gas to the reaction mixture has no effect on the equilibrium position because the concentrations of the gases are not changed.* This is a common misconception for students.

e) *If the volume of the reaction vessel is changed, the concentrations of the gases are changed because for gases, M $\propto$ n/V. If the vessel's volume is decreased, the equilibrium position will shift to the side that has the fewest moles of gas.* In this exercise the right or product side has the fewest moles.

f) *Adding a catalyst has no effect on the position of equilibrium.* Catalysts change the rates of reactions but not positions of equilibrium.

The Reaction Quotient, Q, Sample Exercise

6. *A nonequilibrium mixture has a $[NO_2] = 0.50$ M and $[N_2O_4] = 0.50$ M at 298 K. How will this reaction respond as equilibrium is reestablished? Will the concentration of the reactants increase or decrease? Will the concentration of the products increase or decrease?*

$$2\ NO_{2(g)} \rightleftarrows N_2O_{4(g)}$$

The correct answer is the concentration of the products will increase and the concentration of the reactants will decrease until equilibrium is reestablished.

INSIGHT: The reaction quotient, Q, is used to predict how nonequilibrium mixtures will respond as they reestablish equilibrium. Remember, Q uses the nonequilibrium concentrations whereas K_c uses equilibrium concentrations.

$$Q = \frac{[N_2O_4]}{[NO_2]^2} = \frac{[0.50]}{[0.50]^2} = 2.0 \text{ and thus } Q < K_c$$

In exercise 4, we determined that $K_c = 6.84$ for this reaction, thus $Q < K_c$. Because Q is smaller than K_c, this implies that the mixture has too few products and too many reactants. (Q is a fraction and its size is telling us that the numerator needs to be bigger and the denominator smaller to return to equilibrium.) Thus the concentration of the products will increase and the concentration of the reactants will decrease until equilibrium is reestablished.

1. If $Q < K_c$, the reaction will consume reactants and yield products to reestablish equilibrium.
2. If $Q > K_c$, the reaction will produce reactants and consume products to reestablish equilibrium.
3. If $Q = K_c$, the reaction is at equilibrium.

Relationship of ΔG^0_{rxn} to the Equilibrium Constant Sample Exercise

7. ***What is the value of the gaseous equilibrium constant, K_P, at 298 K for this reaction?***

$$H_{2(g)} + F_{2(g)} \rightleftarrows 2\,HF_{(g)}$$

The correct answer is 5.11×10^{95}.

We can calculate ΔG^0_{rxn} using the method described in Module 15. For this reaction $\Delta G^0_{rxn} = -546$ kJ/mol or -5.46×10^5 J/mol.

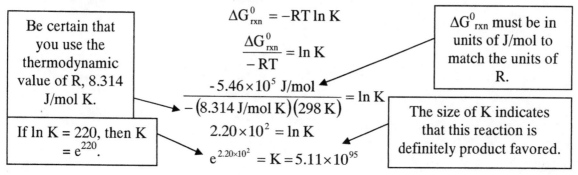

Be certain that you use the thermodynamic value of R, 8.314 J/mol K.

If ln K = 220, then K = e^{220}.

$$\Delta G^0_{rxn} = -RT \ln K$$

$$\frac{\Delta G^0_{rxn}}{-RT} = \ln K$$

$$\frac{-5.46 \times 10^5 \text{ J/mol}}{-(8.314 \text{ J/mol K})(298 \text{ K})} = \ln K$$

$$2.20 \times 10^2 = \ln K$$

$$e^{2.20 \times 10^2} = K = 5.11 \times 10^{95}$$

ΔG^0_{rxn} must be in units of J/mol to match the units of R.

The size of K indicates that this reaction is definitely product favored.

Evaluation of an Equilibrium Constant at a Different Temperature Sample Exercise

8. ***The following reaction has an equilibrium constant, K_c, of 6.84 at 298 K. What is the value of K_c at 225K?***

$$2\,NO_{2(g)} \rightleftarrows N_2O_{4(g)}$$

The correct answer is 1.15×10^4.

The problem tells us that $K_{T_1} = 6.84$, $T_1 = 298$K, and $T_2 = 225$K. From Module 15 we can calculate that $\Delta H^0 = -57.2$ kJ/mol $= -5.72 \times 10^4$ J/mol.

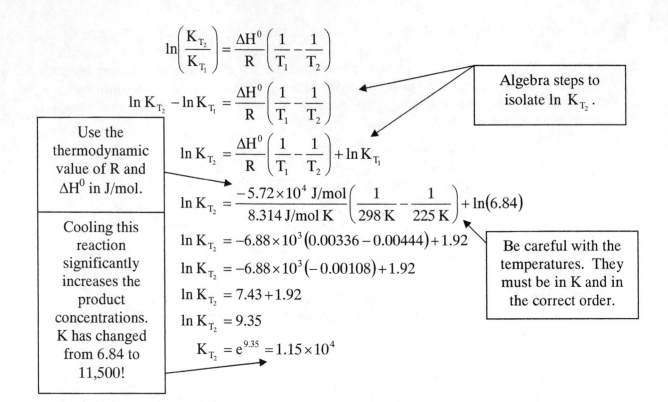

Module 18
Aqueous Equilibria

Introduction

This module describes the calculations required to determine the concentrations of species in various types of aqueous solutions. Exercises in this module will include determining the hydronium and hydroxide concentrations in solutions; the pH, pOH, and % ionization of various solutions; acid-base titration calculations, the pH and pOH of solutions in hydrolysis calculations; the concentrations and pH of buffer solutions; and the ion concentrations for insoluble solids. These calculations are relatively simple to perform but easy to get confused.

Module 18 Key Equations & Concepts

1. $K_w = [H_3O^+][OH^-] = 1.00 \times 10^{-14}$

   $14 = pH + pOH$

   This is the ionization constant for water, K_w. It is used to calculate either the hydronium or hydroxide ion concentration in aqueous solutions given the concentration of either one of these ions. The second equation is mathematically equivalent to the first one. It relates the pH to the pOH and vice versa.

   $pH = -\log[H^+]$

2. $pOH = -\log[OH^-]$

   $pK_a = -\log K_a$

   In chemistry, the symbol pX is defined as the $-\log X$. In these equations the pH, pOH, and pK_a are defined. pH is a condensed method to write the H^+ or H_3O^+ concentration in aqueous solutions. pOH is an equivalent method of writing the aqueous OH^- concentration. pK_a is the shorthand method for writing Ka values.

   For the weak acid equilibrium $HA \rightleftarrows H^+ + A^-$ $K_a = \dfrac{[H^+][A^-]}{[HA]}$

3. For the weak base equilibrium $BOH \rightleftarrows B^+ + OH^-$ $K_b = \dfrac{[B^+][OH^-]}{BOH}$

   $\% \text{ ionization} = \dfrac{[\text{ionized species}]}{[\text{initial species}]} \times 100$

   K_a and K_b are ionization constants for weak acids and weak bases, respectively. They are used to determine the concentrations of all species in aqueous solutions of weak acids and bases. The % ionization is also used in weak acid and base calculations to describe how much of the acid or base is ionized in solution. K_a and K_b values are tabulated in an appendix in your textbook.

4. For an acidic buffer solution $pH = pK_a + \log \dfrac{[\text{salt}]}{[\text{acid}]}$

   For a basic buffer solution $pOH = pK_b + \log \dfrac{[\text{salt}]}{[\text{base}]}$

The Henderson-Hasselbalch equations are used to find the pH of buffer solutions given concentrations of the salt and acid for acidic buffers, or the salt and base for basic buffers.

5. For acid - base conjugate pairs $K_w = K_a \times K_b$

In buffer and hydrolysis calculations, this relationship is used to determine the acid ionization constant for the conjugate acid of a weak base or the base ionization constant for the conjugate base of a weak acid.

6. For a water insoluble solid where $M_yX_{z(s)} \rightleftharpoons yM^{Z+}_{(aq)} + zX^{Y-}_{(aq)}$ $K_{sp} = \left[M^{Z+}\right]^y \left[X^{Y-}\right]^z$

The solubility product constant, K_{sp}, describes the ion concentrations of insoluble solids in aqueous solutions.

Water Ionization Constant Sample Exercises

1. **What is the [OH⁻] in an aqueous solution that has a pH of 5.25?**
   The correct answer is $[OH^-] = 1.8 \times 10^{-9}$ M.

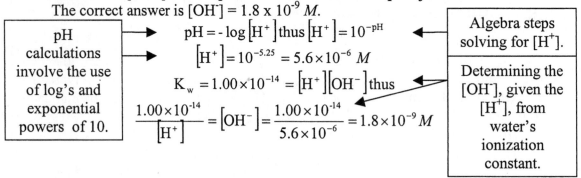

pH calculations involve the use of log's and exponential powers of 10.

$$pH = -\log\left[H^+\right] \text{ thus } \left[H^+\right] = 10^{-pH}$$
$$\left[H^+\right] = 10^{-5.25} = 5.6 \times 10^{-6} \ M$$
$$K_w = 1.00 \times 10^{-14} = \left[H^+\right]\left[OH^-\right] \text{ thus}$$
$$\frac{1.00 \times 10^{-14}}{\left[H^+\right]} = \left[OH^-\right] = \frac{1.00 \times 10^{-14}}{5.6 \times 10^{-6}} = 1.8 \times 10^{-9} \ M$$

Algebra steps solving for $[H^+]$.

Determining the [OH⁻], given the [H⁺], from water's ionization constant.

2. **What is the pH of an aqueous solution that has a [OH⁻] = 3.45 x 10⁻³?**
   The correct answer is pH = 11.538.

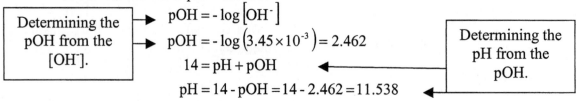

Determining the pOH from the [OH⁻].

$$pOH = -\log\left[OH^-\right]$$
$$pOH = -\log\left(3.45 \times 10^{-3}\right) = 2.462$$
$$14 = pH + pOH$$
$$pH = 14 - pOH = 14 - 2.462 = 11.538$$

Determining the pH from the pOH.

Strong Acid or Base Dissociation Sample Exercise

3. **What is the pH of an aqueous 0.025 M Sr(OH)₂ solution?**
   The correct answer is pH = 12.70.

YIELD

In all aqueous equilibrium problems, you must first decide if the solution is a strong acid/base equilibrium or a weak/acid base equilibrium. Strong acid/base equilibria are easiest to calculate because the acids and bases ionize ~ 100% in water. Watch for polyprotic acids and polyhydroxy bases as they will have increased ion concentrations.

INSIGHT:

Strontium hydroxide, Sr(OH)₂, is a water soluble, strong, polyhydroxy base. When it dissociates in water, the [OH⁻] in solution will be twice the molarity of the Sr(OH)₂. This will also be true for Ca(OH)₂ and Ba(OH)₂. There is one strong polyprotic acid that you need to be aware of, H₂SO₄. For H₂SO₄, the [H⁺] will be twice the molarity.

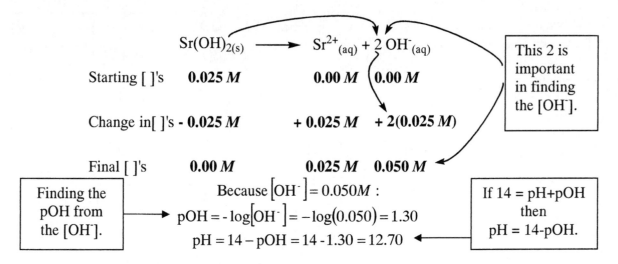

	$Sr(OH)_{2(s)} \longrightarrow$	$Sr^{2+}_{(aq)} + 2\ OH^-_{(aq)}$	
Starting [ ]'s	**0.025 M**	**0.00 M** **0.00 M**	This 2 is important in finding the [OH⁻].
Change in[ ]'s	**- 0.025 M**	**+ 0.025 M** **+ 2(0.025 M)**	
Final [ ]'s	**0.00 M**	**0.025 M** **0.050 M**	

Finding the pOH from the [OH⁻].

Because $\left[OH^-\right] = 0.050M$:

$$pOH = -\log\left[OH^-\right] = -\log(0.050) = 1.30$$
$$pH = 14 - pOH = 14 - 1.30 = 12.70$$

If 14 = pH+pOH then pH = 14-pOH.

4. How many mL of 0.125 M HCl are required to exactly neutralize 25.0 mL of an aqueous 0.025 M Sr(OH)₂ solution?
The correct answer is 10.0 mL.

INSIGHT:

The words "exactly neutralize" or "neutralize" are your clue that this is a titration problem. You should also note that it is the reaction of a strong acid, HCl, with the dihydroxy strong base, Sr(OH)₂. In all titrations, the 1st step must be to **write a balanced chemical reaction**.

Important balanced reaction.

$$2\,HCl_{(aq)} + Sr(OH)_{2(aq)} \rightarrow SrCl_{2(aq)} + 2H_2O_{(\ell)}$$

$$?\ mmol\ Sr(OH)_2 = (25.0\ mL\ HCl)(0.025M\ HCl) = 0.625\ mmol\ Sr(OH)_2$$

$$?\ mL\ Sr(OH)_2 = (0.625\ mmol\ Sr(OH)_2)\left(\frac{2\ mmol\ HCl}{1\ mmol\ Sr(OH)_2}\right)\left(\frac{1\ mL\ HCl}{0.125\ mmol\ HCl}\right) = 10.0\ mL$$

This reaction ratio is important.

M inverted and used as a conversion factor.

Weak Acid or Base Ionization Sample Exercise

5. What is the pH and % ionization of an aqueous 0.125 M acetic acid, CH₃COOH, solution? $K_a = 1.8 \times 10^{-5}$
The correct answer is pH = 2.82 and % ionization =1.2%.

INSIGHT:

Key clues to recognizing a weak acid/base equilibrium problem are:
a) The chemical compound ionizing is a weak acid or base.
b) Presence of a K_a or K_b value.
c) Question will ask for the pH, pOH, % ionization or a combination.

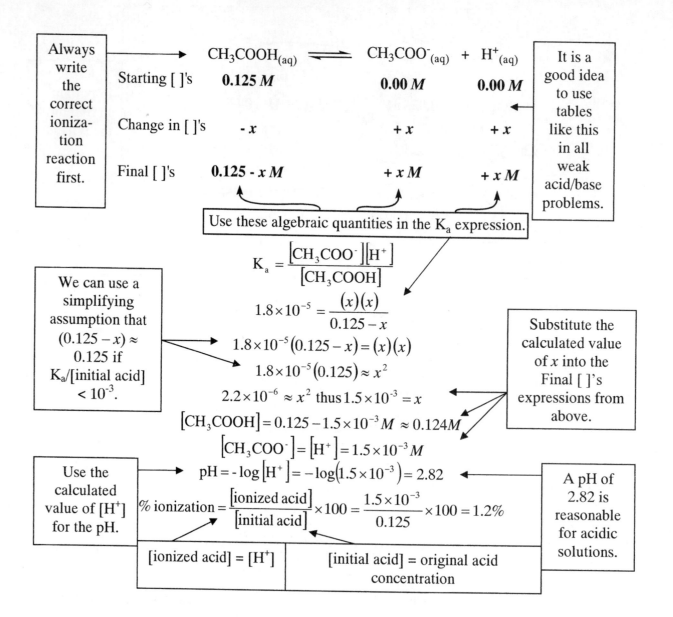

Always write the correct ionization reaction first. →

$$CH_3COOH_{(aq)} \rightleftharpoons CH_3COO^-_{(aq)} + H^+_{(aq)}$$

	Starting [ ]'s	0.125 M	0.00 M	0.00 M
	Change in [ ]'s	- x	+ x	+ x
	Final [ ]'s	0.125 - x M	+ x M	+ x M

It is a good idea to use tables like this in all weak acid/base problems.

Use these algebraic quantities in the K_a expression.

$$K_a = \frac{[CH_3COO^-][H^+]}{[CH_3COOH]}$$

$$1.8\times10^{-5} = \frac{(x)(x)}{0.125-x}$$

We can use a simplifying assumption that $(0.125-x) \approx 0.125$ if $K_a/[\text{initial acid}] < 10^{-3}$.

$$1.8\times10^{-5}(0.125-x)=(x)(x)$$

$$1.8\times10^{-5}(0.125) \approx x^2$$

$$2.2\times10^{-6} \approx x^2 \text{ thus } 1.5\times10^{-3} = x$$

Substitute the calculated value of x into the Final [ ]'s expressions from above.

$$[CH_3COOH]=0.125-1.5\times10^{-3}M \approx 0.124M$$

$$[CH_3COO^-]=[H^+]=1.5\times10^{-3}M$$

Use the calculated value of $[H^+]$ for the pH. →

$$pH = -\log[H^+] = -\log(1.5\times10^{-3}) = 2.82$$

$$\%\ \text{ionization} = \frac{[\text{ionized acid}]}{[\text{initial acid}]}\times100 = \frac{1.5\times10^{-3}}{0.125}\times100 = 1.2\%$$

A pH of 2.82 is reasonable for acidic solutions.

[ionized acid] = [H⁺]

[initial acid] = original acid concentration

Hydrolysis or Solvolysis Solution Sample Exercise

6. *What are the pH and pOH of an aqueous 0.125 M sodium acetate, NaCH₃COO, solution? $K_a = 1.8 \times 10^{-5}$*

   The correct answer is pH = 8.92 and pOH =5.08.

INSIGHT:

Key clues to recognizing a hydrolysis or solvolysis problem:
   a) The chemical compound ionizing is a soluble salt of a weak acid or base. (NaCH₃COO is the soluble salt of acetic acid.)
   b) A K_a or K_b will be provided but the salt contains the conjugate base or conjugate acid. You will have to use $K_w = K_a \times K_b$ to get the required ionization constant.
   c) **If the salt contains the anion of a weak acid, the solution will be basic.** You will need the K_b for the ionization calculation.

d) **If the salt contains the cation of a weak base, the solution will be acidic.** You will need the K_a for the ionization calculation.

e) The ionization equation will involve the reaction of the salt with water. One of the ions will be a spectator ion which can be ignored in the ionization equation. (In this exercise, Na^+ is the spectator ion.)

f) Questions will commonly ask for pH or pOH of the solution.

g) The simplifying assumption used in exercise 4 usually will be applicable in these problems as the ionized concentrations are small.

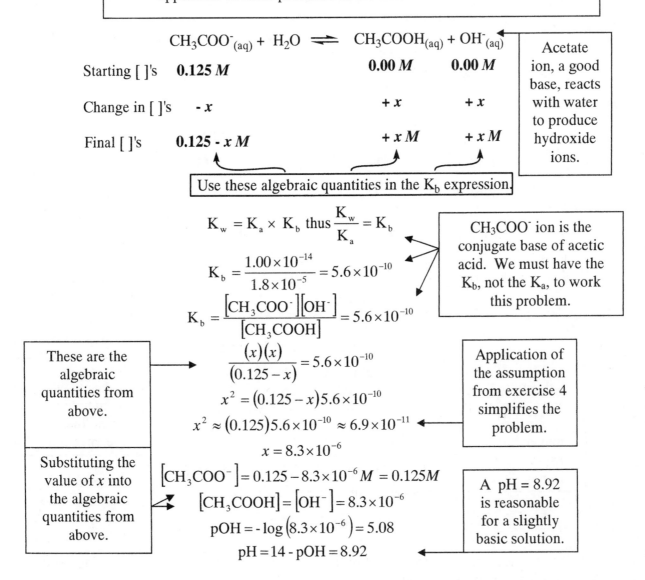

$$CH_3COO^-_{(aq)} + H_2O \rightleftharpoons CH_3COOH_{(aq)} + OH^-_{(aq)}$$

Starting [ ]'s	0.125 M	0.00 M	0.00 M
Change in [ ]'s	- x	+ x	+ x
Final [ ]'s	0.125 - x M	+ x M	+ x M

Acetate ion, a good base, reacts with water to produce hydroxide ions.

Use these algebraic quantities in the K_b expression.

$$K_w = K_a \times K_b \text{ thus } \frac{K_w}{K_a} = K_b$$

CH_3COO^- ion is the conjugate base of acetic acid. We must have the K_b, not the K_a, to work this problem.

$$K_b = \frac{1.00 \times 10^{-14}}{1.8 \times 10^{-5}} = 5.6 \times 10^{-10}$$

$$K_b = \frac{[CH_3COO^-][OH^-]}{[CH_3COOH]} = 5.6 \times 10^{-10}$$

These are the algebraic quantities from above.

$$\frac{(x)(x)}{(0.125 - x)} = 5.6 \times 10^{-10}$$

$$x^2 = (0.125 - x)5.6 \times 10^{-10}$$

$$x^2 \approx (0.125)5.6 \times 10^{-10} \approx 6.9 \times 10^{-11}$$

Application of the assumption from exercise 4 simplifies the problem.

$$x = 8.3 \times 10^{-6}$$

Substituting the value of x into the algebraic quantities from above.

$$[CH_3COO^-] = 0.125 - 8.3 \times 10^{-6} M = 0.125 M$$

$$[CH_3COOH] = [OH^-] = 8.3 \times 10^{-6}$$

$$pOH = -\log(8.3 \times 10^{-6}) = 5.08$$

$$pH = 14 - pOH = 8.92$$

A pH = 8.92 is reasonable for a slightly basic solution.

Buffer Solution Sample Exercise

7. *What are the concentrations of the relevant species and the pH of a solution that is 0.100 M in acetic acid, CH_3COOH, and 0.025 M in sodium acetate, $NaCH_3COO$? $K_a = 1.8 \times 10^{-5}$*

The correct answer is $[CH_3COOH] = 0.100 M$, $[CH_3COO^-] = 0.025 M$, $[H^+] = 7.2 \times 10^{-5} M$, and pH = 4.14.

Key clues to recognizing a buffer problem:
  a) The solution will contain a soluble salt dissolved in either a weak acid or a weak base. Concentrations or amounts of both will be present in the problem.
  b) The salt must be the conjugate partner of the weak acid or weak base.
  c) The equilibrium table will have starting concentrations of both the salt and the acid or base. The equilibrium will involve the common ion effect.
  d) Henderson-Hasselbalch equations are a simple method to find the buffer solution's pH.
  e) Simplifying assumption will be useful to quickly solve the solution concentration problem.

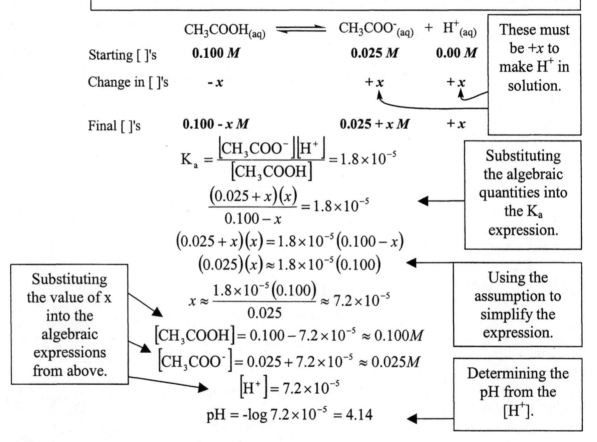

$$CH_3COOH_{(aq)} \rightleftharpoons CH_3COO^-_{(aq)} + H^+_{(aq)}$$

These must be $+x$ to make H^+ in solution.

Starting [ ]'s   0.100 M 0.025 M 0.00 M

Change in [ ]'s $-x$ $+x$ $+x$

Final [ ]'s   0.100 $-x$ M 0.025 $+x$ M $+x$

$$K_a = \frac{[CH_3COO^-][H^+]}{[CH_3COOH]} = 1.8 \times 10^{-5}$$

Substituting the algebraic quantities into the K_a expression.

$$\frac{(0.025 + x)(x)}{0.100 - x} = 1.8 \times 10^{-5}$$

$$(0.025 + x)(x) = 1.8 \times 10^{-5}(0.100 - x)$$

$$(0.025)(x) \approx 1.8 \times 10^{-5}(0.100)$$

Using the assumption to simplify the expression.

Substituting the value of x into the algebraic expressions from above.

$$x \approx \frac{1.8 \times 10^{-5}(0.100)}{0.025} \approx 7.2 \times 10^{-5}$$

$$[CH_3COOH] = 0.100 - 7.2 \times 10^{-5} \approx 0.100 M$$

$$[CH_3COO^-] = 0.025 + 7.2 \times 10^{-5} \approx 0.025 M$$

$$[H^+] = 7.2 \times 10^{-5}$$

$$pH = -\log 7.2 \times 10^{-5} = 4.14$$

Determining the pH from the $[H^+]$.

If the question only asks for the pH of the buffer solution, the simplest method to get that answer is to use the Henderson-Hasselbalch equations.

Solubility Product Sample Exercises

8. *The solubility of iron(II) hydroxide, Fe(OH)$_2$, in water is 1.1 x 10^3 g/L at 25.0°C. What is the solubility product constant for Fe(OH)$_2$?*
   The correct answer is $K_{sp} = 6.9 \times 10^{-15}$.

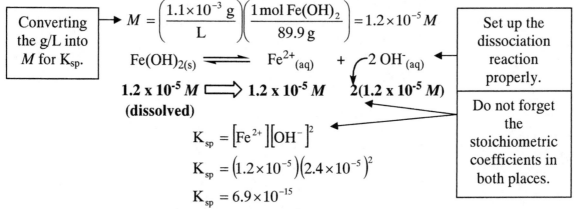

Converting the g/L into M for K_{sp}.

$$M = \left(\frac{1.1 \times 10^{-3} \text{ g}}{L}\right)\left(\frac{1 \text{ mol Fe(OH)}_2}{89.9 \text{ g}}\right) = 1.2 \times 10^{-5} M$$

$$Fe(OH)_{2(s)} \rightleftharpoons Fe^{2+}_{(aq)} + 2\,OH^-_{(aq)}$$

$1.2 \times 10^{-5}\,M \implies 1.2 \times 10^{-5}\,M \quad 2(1.2 \times 10^{-5}\,M)$

(dissolved)

Set up the dissociation reaction properly.

Do not forget the stoichiometric coefficients in both places.

$$K_{sp} = \left[Fe^{2+}\right]\left[OH^-\right]^2$$

$$K_{sp} = \left(1.2 \times 10^{-5}\right)\left(2.4 \times 10^{-5}\right)^2$$

$$K_{sp} = 6.9 \times 10^{-15}$$

9. *What are the molar solubilities of Zn^{2+} and OH^- for zinc hydroxide at $25.0^\circ C$?*
$K_{sp} = 4.5 \times 10^{-17}$

The correct answer is $[Zn^{2+}] = 2.2 \times 10^{-5}\,M$, $[OH^-] = 4.4 \times 10^{-5}\,M$.

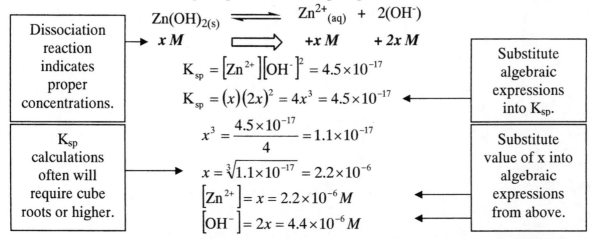

$$Zn(OH)_{2(s)} \rightleftharpoons Zn^{2+}_{(aq)} + 2(OH^-)$$

Dissociation reaction indicates proper concentrations.

$x\,M \implies +x\,M \quad + 2x\,M$

$$K_{sp} = \left[Zn^{2+}\right]\left[OH^-\right]^2 = 4.5 \times 10^{-17}$$

$$K_{sp} = (x)(2x)^2 = 4x^3 = 4.5 \times 10^{-17}$$

Substitute algebraic expressions into K_{sp}.

$$x^3 = \frac{4.5 \times 10^{-17}}{4} = 1.1 \times 10^{-17}$$

K_{sp} calculations often will require cube roots or higher.

$$x = \sqrt[3]{1.1 \times 10^{-17}} = 2.2 \times 10^{-6}$$

$$\left[Zn^{2+}\right] = x = 2.2 \times 10^{-6}\,M$$

$$\left[OH^-\right] = 2x = 4.4 \times 10^{-6}\,M$$

Substitute value of x into algebraic expressions from above.

YIELD

Tips on aqueous equilibrium problems:
a) The hardest part is deciding if the problem is a weak acid/base, a solvolysis, a buffer, or a solubility product problem. The **INSIGHT** columns provide the key clues in determining which problem you are working on. Be very familiar with these.
b) Always write the correct ionization reaction and ionization expression (K_a, K_b, K_{sp}, etc.) for the problem you are given then set up the appropriate table underneath the ionization reaction to help with the algebra.
c) Correct use of the simplifying assumption will save enormous amounts of time and yield the correct answer. Use it whenever you can!

d) Every one of these equilibrium problems uses the same basic mathematical method to determine the concentrations and pH, etc. The only differences are setting up the ionization reactions and placing concentrations in the appropriate places in the table. If you can recognize the problem type, solving the problem is very straightforward after that.

e) The hardest equilibrium for students to recognize is always the hydrolysis problems. Keep an eye out for these.

Module 19
Electrochemistry

Introduction

This module describes the basic electrochemistry methods used in typical textbooks. In this module we will look at how to: balance redox reactions in acidic and basic solutions; the differences between electrolytic and voltaic cells; use Faraday's law to determine the amount of a species that is reduced in an electrolytic cell; determine the anode, cathode, and electron flow in both cell types; determine the standard cell potential for a voltaic cell; use the Nernst equation to find the cell potential at nonstandard conditions; and determine the Gibb's Free Energy change and equilibrium constant for a cell from its standard potential. You will need your textbook opened to the appendix containing the standard electrode potentials to understand this material.

Module 19 Key Equations & Concepts

1. **Electrolytic cells are electrochemical cells in which nonspontaneous chemical reactions are forced to occur by the application of an external voltage.**

    In electrolytic cells reactions that would not occur in nature, such as the electrolysis of chemical compounds or the electroplating of metals, are made to occur by the passage of electricity through the cell.

2. **Voltaic cells are electrolytic cells in which spontaneous chemical reactions occur and the electrons generated in the reaction are passed through an external wire.**

    Voltaic cells are batteries such as dry cells and lithium batteries used in watches, cameras, etc.

3. **The anode is the electrode where oxidation occurs in both electrolytic and voltaic cells. The cathode is the electrode where reduction occurs in both cell types.**

    In electrolytic cells the anode is the positive electrode and the cathode is the negative electrode. This is reversed for voltaic cells where the anode is the negative electrode and the cathode is the positive electrode.

4. **Faraday's law of electrolysis states that the amount of a chemical compound oxidized or reduced at an electrode during electrolysis is directly proportional to the amount of electricity passed through the cell.**

    This law is used to calculate the number of grams of a chemical compound transformed from oxidized to reduced species, or vice versa, in an electrolytic cell. Faraday's constant, 1 faraday = 9.65×10^4 coulombs, is essential in these calculations.

5. **Standard Cell Potentials are the initial voltage produced in a voltaic cell at standard conditions.**

    To find the standard cell potential, add the standard cell potential for the reduction step to the reverse of the standard cell potential for the oxidation step. Standard cell potentials are tabulated in your textbook in the appendices.

6. $E = E^0 - \dfrac{2.303\,RT}{n\,F} \log Q$ or $E = E^0 - \dfrac{0.0592}{n} \log Q$ at $25.0^\circ C$

The Nernst equation is used to calculate a cell's potential at nonstandard conditions. E is the nonstandard cell potential, E^0 is the standard cell potential, n is the number of moles of electrons in the reaction, F is Faraday's constant, and Q is the reaction quotient.

7. $\Delta G^0 = -nFE^0_{cell}$ or $nFE^0_{cell} = RT \ln K$

This equation is used to determine either the Gibbs Free Energy change or the equilibrium constant of a chemical reaction once the cell potential has been determined.

Balancing Redox Reactions Sample Exercises

1. Balance the following redox reaction in acidic solution.

$$Cu_{(s)} + NO_{3\ (aq)}^{-} \rightarrow Cu^{2+}_{\ (aq)} + NO_{2(g)}$$

The correct answer is $Cu + 4\,H^+ + 2\,NO_3^- \rightarrow Cu^{2+} + 2\,NO_2 + 2\,H_2O$.

INSIGHT:	Redox reactions can occur in either acidic or basic solutions. In acidic solutions you can add H^+ and H_2O to balance the reaction. In basic solutions you can add OH^- and H_2O. The problem will either state that the reaction is in acidic solution or H^+ will be present as a reactant or product. Similarly, for basic solutions look for a statement that the reaction is in basic solution or the presence of OH^-.
YIELD	There are two simple methods to balance redox reactions, the change in oxidation number method and the half-reaction method. The change in oxidation number method is more physically correct but the half-reaction method is simpler to learn and more straight forward. We will use the half-reaction method in these exercises. It is very important that you review the rules for assigning oxidation numbers and for balancing redox reactions found in your textbook.

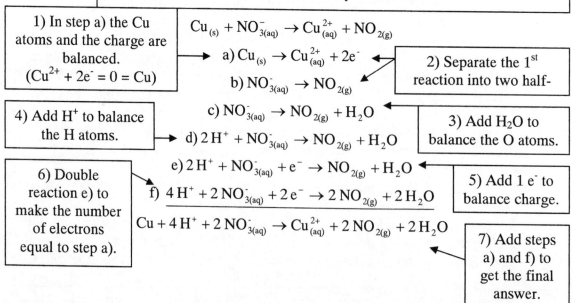

104

2. Balance the following redox reaction in basic solution.

$$CrO_2^-{}_{(aq)} + ClO^-{}_{(aq)} \rightarrow CrO_4^-{}_{(aq)} + Cl^-{}_{(aq)}$$

The correct answer is $CrO_2^- + 2ClO^- \rightarrow CrO_4^- + 2Cl^-$.

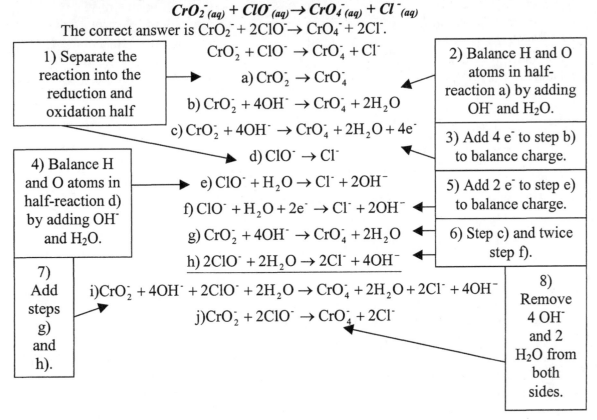

1) Separate the reaction into the reduction and oxidation half

$CrO_2^- + ClO^- \rightarrow CrO_4^- + Cl^-$

a) $CrO_2^- \rightarrow CrO_4^-$

b) $CrO_2^- + 4OH^- \rightarrow CrO_4^- + 2H_2O$

c) $CrO_2^- + 4OH^- \rightarrow CrO_4^- + 2H_2O + 4e^-$

d) $ClO^- \rightarrow Cl^-$

e) $ClO^- + H_2O \rightarrow Cl^- + 2OH^-$

f) $ClO^- + H_2O + 2e^- \rightarrow Cl^- + 2OH^-$

g) $CrO_2^- + 4OH^- \rightarrow CrO_4^- + 2H_2O$

h) $2ClO^- + 2H_2O \rightarrow 2Cl^- + 4OH^-$

i) $CrO_2^- + 4OH^- + 2ClO^- + 2H_2O \rightarrow CrO_4^- + 2H_2O + 2Cl^- + 4OH^-$

j) $CrO_2^- + 2ClO^- \rightarrow CrO_4^- + 2Cl^-$

2) Balance H and O atoms in half-reaction a) by adding OH^- and H_2O.

3) Add 4 e⁻ to step b) to balance charge.

4) Balance H and O atoms in half-reaction d) by adding OH^- and H_2O.

5) Add 2 e⁻ to step e) to balance charge.

6) Step c) and twice step f).

7) Add steps g) and h).

8) Remove 4 OH⁻ and 2 H₂O from both sides.

Electrolytic Cell Sample Exercise

3. **An electrolytic cell containing an aqueous NaCl solution is constructed. What chemical species will be produced at the cathode and anode? What is the direction of electron flow in this cell?**

The correct answer is $H_{2(g)}$ is produced at the cathode and $Cl_{2(g)}$ is produced at the anode. The electrons flow from the anode passing through the battery and to the cathode.

INSIGHT: In this electrolytic cell there are four possible redox reactions. The two possible reductions are $Na^+{}_{(aq)}$ to $Na_{(s)}$ or H_2O to $H_{2(g)}$ and $OH^-{}_{(aq)}$. The two possible oxidations are $Cl^-{}_{(aq)}$ to $Cl_{2(g)}$ or H_2O to $H^+{}_{(aq)}$ and O_2. *How do you determine the correct reactions in electrolytic cells?* Electrolytic cells force nonspontaneous chemical reactions to occur. The standard reduction potentials in your textbook will tell you which one to choose. 1) The oxidation reaction will be the one of the two possible reactions that has the most positive reduction potential (most negative oxidation potential). 2) The reduction reaction will be the one of the two possible reactions that has the most positive, or least negative, reduction potential.

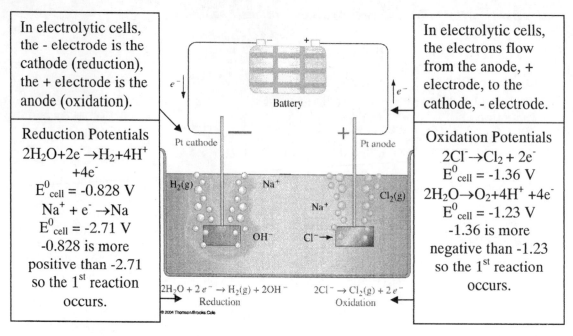

In electrolytic cells, the - electrode is the cathode (reduction), the + electrode is the anode (oxidation).		In electrolytic cells, the electrons flow from the anode, + electrode, to the cathode, - electrode.
Reduction Potentials $2H_2O+2e^-\rightarrow H_2+4H^+$ $+4e^-$ $E^0_{cell} = -0.828$ V $Na^+ + e^- \rightarrow Na$ $E^0_{cell} = -2.71$ V -0.828 is more positive than -2.71 so the 1st reaction occurs.		**Oxidation Potentials** $2Cl^-\rightarrow Cl_2 + 2e^-$ $E^0_{cell} = -1.36$ V $2H_2O\rightarrow O_2+4H^+ +4e^-$ $E^0_{cell} = -1.23$ V -1.36 is more negative than -1.23 so the 1st reaction occurs.

Faraday's Law Sample Exercise

4. *How many grams of nickel metal will be produced at the cathode when 3.75 amps of current are passed for 75.0 minutes through an electrolytic cell containing NiSO$_{4(aq)}$?*

    The correct answer is 5.15 g of Ni.

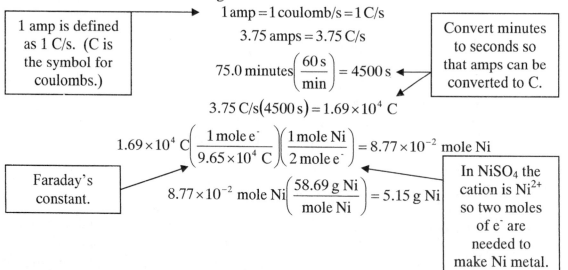

1 amp is defined as 1 C/s. (C is the symbol for coulombs.)	$1\,amp = 1\,coulomb/s = 1\,C/s$ $3.75\,amps = 3.75\,C/s$ $75.0\,minutes\left(\dfrac{60\,s}{min}\right) = 4500\,s$	Convert minutes to seconds so that amps can be converted to C.

$$3.75\,C/s(4500\,s) = 1.69\times10^4\,C$$

Faraday's constant.	$1.69\times10^4\,C\left(\dfrac{1\,mole\,e^-}{9.65\times10^4\,C}\right)\left(\dfrac{1\,mole\,Ni}{2\,mole\,e^-}\right) = 8.77\times10^{-2}\,mole\,Ni$ $8.77\times10^{-2}\,mole\,Ni\left(\dfrac{58.69\,g\,Ni}{mole\,Ni}\right) = 5.15\,g\,Ni$	In NiSO$_4$ the cation is Ni^{2+} so two moles of e$^-$ are needed to make Ni metal.

Voltaic Cell Sample Exercises

5. *A voltaic cell is constructed of a 1.0 M CuSO$_{4(aq)}$ solution and a 1.0 M AgNO$_{3(aq)}$ solution plus the electrodes, connecting wires, and salt bridges. What chemical species will be produced at the cathode and anode? What is the direction of electron flow in this cell?*

    The correct answer is Cu is oxidized to Cu^{2+} at the anode and Ag$^+$ is reduced to Ag at the cathode. The electrons flow from the anode to the cathode in this cell. Because the reactions are spontaneous, no battery is needed.

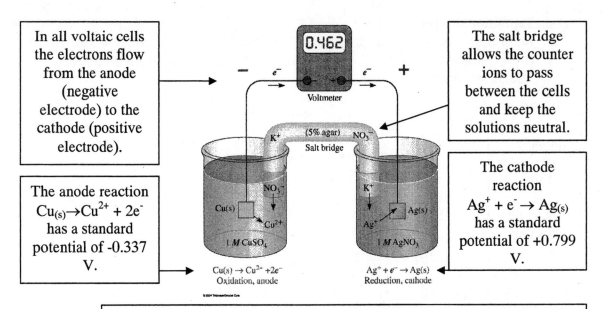

| In all voltaic cells the electrons flow from the anode (negative electrode) to the cathode (positive electrode). | The salt bridge allows the counter ions to pass between the cells and keep the solutions neutral. |

| The anode reaction $Cu_{(s)} \rightarrow Cu^{2+} + 2e^-$ has a standard potential of -0.337 V. | The cathode reaction $Ag^+ + e^- \rightarrow Ag_{(s)}$ has a standard potential of +0.799 V. |

INSIGHT: In voltaic cells the reactions are spontaneous. You can predict the reactions using the standard reduction potentials. 1) The oxidation reaction will have the least positive (most negative) standard reduction potential. 2) The reduction reaction will have the most positive (least negative) standard reduction potential.

6. *What is the standard cell potential for the voltaic cell in Sample Exercise 5?*
The correct answer is 0.462 V.

| The reduction reaction is balanced **but the E^0_{cell} is not doubled.** Cell potentials are intensive quantities. | $2\,Ag^+_{(aq)} + 2\,e^- \rightarrow 2\,Ag_{(s)}$ $E^0 = +0.799\,V$
 $Cu_{(s)} \rightarrow Cu^{2+}_{(aq)} + 2\,e^-$ $E^0 = -0.337\,V$
 _____
 $2\,Ag^+_{(aq)} + Cu_{(s)} \rightarrow 2\,Ag_{(s)} + Cu^{2+}_{(aq)}$ $E^0_{cell} = +0.462\,V$

 The cell potentials, E^0, are from the tables in the appendix of your textbook. In voltaic cells, the overall cell potential E^0_{cell} must be a <u>positive</u> value. | Always reverse the reaction and change the sign of the tabulated reduction potential for the oxidation. |

YIELD

To calculate the standard cell potential follow these steps:
1) Write the half-reaction and the cell potential for the reaction that has the most positive (or least negative) reduction potential, E^0.
2) Write the half-reaction and the cell potential for the other reaction as an oxidation. To do this you must take the reaction given in the table in your textbook, reverse the reaction and change the sign of E^0.
3) Make sure that the electrons from each half-reaction are balanced but **do not multiply the reduction potentials.**
4) Add the two half-reactions, canceling the electrons, and add the two cell potentials to get the E^0_{cell}. This must always be a positive value to indicate that the reaction is spontaneous.

Nernst Equation Sample Exercise

7. **What is the cell potential for the voltaic cell in Sample Exercise 5 at 325 K if the CuSO₄ concentration is 2.00 M and the AgNO₃ concentration is 3.00 M?**
   The correct answer is 0.483 V.

| E^0 is the cell potential calculated in exercise 4. R is the gas constant. n is the moles of electrons in the reaction. | $E = E^0 - \dfrac{2.303\,RT}{nF}\log Q$

 $E = 0.462\,V - \dfrac{2.303\,(8.314\,\text{J/mol K})(325\,K)}{(2\,\text{mol})(9.65\times10^4\,\text{J/V mol e}^-)}\log\dfrac{2.00}{(3.00)^2}$

 $E = 0.462\,V - \dfrac{6223\,V}{1.93\times10^5}\log\dfrac{2}{9}$

 $E = 0.462\,V - \dfrac{6223\,V}{1.93\times10^5}(-0.653)$

 $E = 0.462\,V - (-0.021\,V) = 0.483\,V$ | F is faraday's constant. Q is the reaction quotient as described in module 17. |

YIELD | The cell potential determined in exercise 3 is at standard conditions (298 K, 1 M solutions, 1 atm pressure, etc.). The Nernst equation allows us to calculate the cell potential at nonstandard conditions such as different temperatures and solution concentrations.

INSIGHT:

1) Because 1 J = V·C thus $(9.65 \times10^4$ C/mol e⁻)(1 J/V·C) = 9.65 x 10⁴ J/V mol e⁻, which is Faraday's constant in a different set of units used in this exercise.

2) This is a heterogeneous equilibrium, there are solids and solutions involved in the reaction. Equilibrium constants, including Q the reaction quotient, for heterogeneous equilibria involve only the species having the largest activities. Solutions have much larger activities than solids and therefore we only use the solution concentrations in Q.

3) The cell reaction is 2 Ag⁺$_{(aq)}$ + Cu$_{(s)}$→ 2 Ag$_{(s)}$ + Cu²⁺$_{(aq)}$. Thus we get

$$Q = \frac{[Cu^{2+}]}{[Ag^+]^2} = \frac{2.00}{(3.00)^2} = \frac{2}{9}.$$

Determination of the Equilibrium Constant for a Voltaic Cell Sample Exercise

8. **What is the equilibrium constant at 298 K for the voltaic cell described in Sample Exercise 5?**
   The correct answer is 4.31 x 10¹⁵.

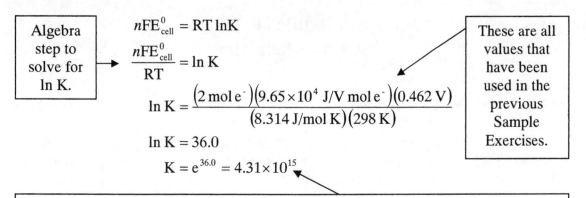

Module 20
Nuclear Chemistry

Introduction

Module 20 discusses the basic relationships used in a typical nuclear chemistry chapter. The important topics described will be how to: <u>calculate the mass defect and binding energy for a nucleus; how to predict the products of alpha, negatron, and positron radioactive decays and the products of a nuclear reaction; and the problems associated with the kinetics of radioactive decay</u>. All of these problems are frequently asked in general chemistry exams.

Module 20 Key Equations & Concepts

1. $\Delta m = [Z(1.0073) + N(1.0087) + Z(0.0005)]$ – **actual mass of atom**

   The mass defect of a nucleus is the difference in the sum of the masses of the protons, neutrons, and electrons in a nucleus minus the actual mass of the atom. This relationship describes how much of the nuclear mass has been converted into energy to bind the nucleus.

2. **Binding Energy** $= \Delta m c^2$

   The nuclear binding energy is the energy required to hold the protons and neutrons together in the nucleus. It is the mass defect converted from mass unit to energy units.

3. $^{A}_{Z}X \rightarrow ^{A-4}_{Z-2}Y + ^{4}_{2}He$

   This is the basic equation for radioactive alpha decay. Alpha decay removes two protons and two neutrons, in the form of a ^{4}He nucleus, from the decaying nucleus converting the element X into a new element Y.

4. $^{A}_{Z}X \rightarrow ^{A}_{Z+1}Y + ^{0}_{-1}e$ (or $^{0}_{-1}\beta^-$)

   Radioactive beta decay, β^- or negatron decay, converts a neutron into a proton by eliminating a high velocity electron, the β^- particle, from the nucleus. The decaying nucleus, X, is converted to a new nucleus, Y, having one additional proton and one less neutron.

5. $^{A}_{Z}X \rightarrow ^{A}_{Z-1}Y + ^{0}_{+1}e$ (or $^{0}_{+1}\beta^+$)

   Radioactive positron decay, β^+, converts a proton into a neutron by eliminating a high velocity positive electron, the β^+ particle, from the nucleus. The decaying nucleus, X is converted to a new nucleus, Y, having one less proton and one more neutron.

6. $^{M_1}_{Z_1}Q \rightarrow ^{M_2}_{Z_2}R + ^{M_3}_{Z_3}Y$ where $M_1 = M_2 + M_3$ and $Z_1 = Z_2 + Z_3$

   This is the basic relationship for nuclear reactions, and radioactive decays. The proton numbers of the product nuclides (Z_2 and Z_3) must sum up to the original nuclide's proton number, Z_1. The mass numbers of the product nuclides (M_2 and M_3) must also add up to the original nuclide's mass, M_1.

7. $A = A_0 e^{-kt}$ and $k\, t_{1/2} = 0.693$

   Radioactive decay obeys first order kinetics as described in Module 16. These are the two important equations for radioactive decay describing the amount of a nuclide remaining after a certain amount of time has passed and the half-life relationship.

Mass Defect Sample Exercise

1. **What is the mass defect, in amu, for a ^{55}Cr nucleus? The actual mass of a ^{55}Cr atom is 54.9408 amu.**

   The correct answer is 0.5161 amu.

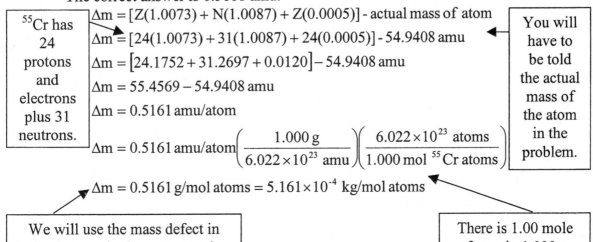

^{55}Cr has 24 protons and electrons plus 31 neutrons.

$\Delta m = [Z(1.0073) + N(1.0087) + Z(0.0005)] - \text{actual mass of atom}$

$\Delta m = [24(1.0073) + 31(1.0087) + 24(0.0005)] - 54.9408 \text{ amu}$

$\Delta m = [24.1752 + 31.2697 + 0.0120] - 54.9408 \text{ amu}$

$\Delta m = 55.4569 - 54.9408 \text{ amu}$

$\Delta m = 0.5161 \text{ amu/atom}$

$\Delta m = 0.5161 \text{ amu/atom} \left(\dfrac{1.000 \text{ g}}{6.022 \times 10^{23} \text{ amu}} \right) \left(\dfrac{6.022 \times 10^{23} \text{ atoms}}{1.000 \text{ mol } ^{55}Cr \text{ atoms}} \right)$

$\Delta m = 0.5161 \text{ g/mol atoms} = 5.161 \times 10^{-4} \text{ kg/mol atoms}$

You will have to be told the actual mass of the atom in the problem.

We will use the mass defect in kg/mol atoms in the next exercise.

There is 1.00 mole of amu in 1.000 g.

INSIGHT: 1.0073 is the mass of a proton, 1.0087 amu is the mass of a neutron, and 0.0005 amu is the mass of an electron.

Binding Energy Sample Exercise

2. **What is the binding energy, in J/mol, for a ^{55}Cr nucleus?**

   The correct answer is 4.65×10^{13} J/mol atoms.

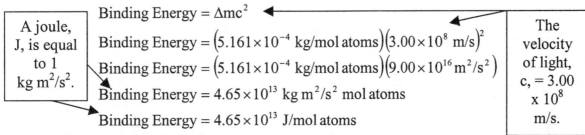

A joule, J, is equal to 1 $kg\,m^2/s^2$.

Binding Energy $= \Delta mc^2$

Binding Energy $= (5.161 \times 10^{-4} \text{ kg/mol atoms})(3.00 \times 10^8 \text{ m/s})^2$

Binding Energy $= (5.161 \times 10^{-4} \text{ kg/mol atoms})(9.00 \times 10^{16} \text{ m}^2/\text{s}^2)$

Binding Energy $= 4.65 \times 10^{13} \text{ kg m}^2/\text{s}^2 \text{ mol atoms}$

Binding Energy $= 4.65 \times 10^{13} \text{ J/mol atoms}$

The velocity of light, c, = 3.00 x 10^8 m/s.

Alpha Decay Sample Exercise

3. **What is the product nuclide of the alpha decay of ^{232}Th?**

   The correct answer is ^{228}Ra.

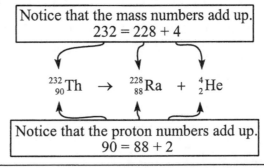

Notice that the mass numbers add up.
$232 = 228 + 4$

$^{232}_{90}Th \rightarrow ^{228}_{88}Ra + ^{4}_{2}He$

Notice that the proton numbers add up.
$90 = 88 + 2$

INSIGHT: Alpha decay occurs primarily in nuclides that have more than 83 protons. To determine the product nuclide, take the proton number of the decaying nucleus and subtract 2. The product's mass number will be the decaying nuclide's mass number minus 4.

Beta Decay Sample Exercises

4. What is the product nuclide of the β^-, negatron, decay of ^{14}C?

The correct answer is ^{14}N.

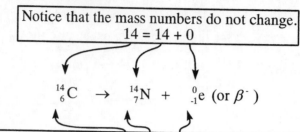

Notice that the mass numbers do not change.
$14 = 14 + 0$

$$^{14}_{6}C \rightarrow \ ^{14}_{7}N \ + \ ^{0}_{-1}e \ (or\ \beta^-)$$

Notice that the charges of the protons and the beta particle add up.
$6 = 7 + (-1)$

5. What is the product nuclide of the β^+, positron, decay of ^{37}Ca?

The correct answer is ^{37}K.

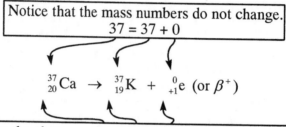

Notice that the mass numbers do not change.
$37 = 37 + 0$

$$^{37}_{20}Ca \rightarrow \ ^{37}_{19}K \ + \ ^{0}_{+1}e \ (or\ \beta^+)$$

Notice that the charges of the protons and the beta particle add up.
$20 = 19 + (+1)$

INSIGHT:
1) In all forms of beta decay, the mass numbers do not change.
2) In β^- decay, the product nuclide will have one **more** proton than the decaying nuclide.
3) In β^+ decay, the product nuclide will have one **less** proton than the decaying nuclide.

Nuclear Reaction Sample Exercise

6. Fill in the missing nuclide in this nuclear reaction.

$$^{53}Cr + \ ^{4}He \rightarrow \underline{\quad} + 2\,n$$

The correct answer is ^{55}Fe.

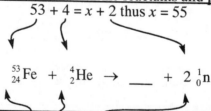

The mass number will be determined from the sum of the mass numbers of the reactants and products.
$53 + 4 = x + 2$ thus $x = 55$

$$^{53}_{24}Fe \ + \ ^{4}_{2}He \rightarrow \underline{\quad} + \ 2\,^{1}_{0}n$$

The proton number will be determined from the sum of the proton numbers of the reactants and products.
$24 + 2 = x + 0$ thus $x = 26$

Fe has 26 protons. The isotope of Fe with a mass of 55 is ^{55}Fe.

Kinetics of Radioactive Decay Sample Exercises

7. *Tritium, ^3H, a radioactive isotope of hydrogen has a half-life of 12.26 y. If 2.0 g of ^3H were made, how much of it would be left 18.40 y later?*
   The correct answer is 0.88 g.

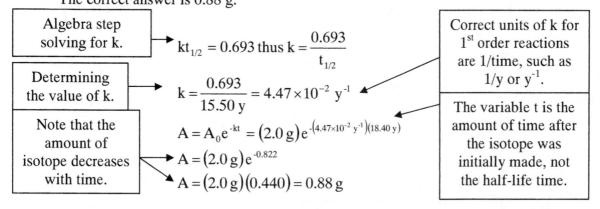

Algebra step solving for k.

$$kt_{1/2} = 0.693 \text{ thus } k = \frac{0.693}{t_{1/2}}$$

Determining the value of k.

$$k = \frac{0.693}{15.50 \text{ y}} = 4.47 \times 10^{-2} \text{ y}^{-1}$$

Note that the amount of isotope decreases with time.

$$A = A_0 e^{-kt} = (2.0 \text{ g}) e^{-(4.47 \times 10^{-2} \text{ y}^{-1})(18.40 \text{ y})}$$
$$A = (2.0 \text{ g}) e^{-0.822}$$
$$A = (2.0 \text{ g})(0.440) = 0.88 \text{ g}$$

Correct units of k for 1st order reactions are 1/time, such as 1/y or y^{-1}.

The variable t is the amount of time after the isotope was initially made, not the half-life time.

8. *A loaf of bread left in the Egyptian temple of Mentukotep II, an ancient pharaoh, has a ^{14}C content that is 61.6% that of living matter. How old is the loaf of bread? You may assume that the decrease in the ^{14}C content is entirely due to the radioactive decay of ^{14}C. The half-life of ^{14}C is 5730 y.*
   The correct answer is 4.00 x 10^3 y.

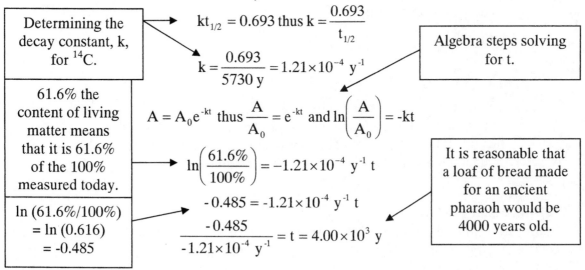

Determining the decay constant, k, for ^{14}C.

$$kt_{1/2} = 0.693 \text{ thus } k = \frac{0.693}{t_{1/2}}$$

$$k = \frac{0.693}{5730 \text{ y}} = 1.21 \times 10^{-4} \text{ y}^{-1}$$

Algebra steps solving for t.

61.6% the content of living matter means that it is 61.6% of the 100% measured today.

$$A = A_0 e^{-kt} \text{ thus } \frac{A}{A_0} = e^{-kt} \text{ and } \ln\left(\frac{A}{A_0}\right) = -kt$$

$$\ln\left(\frac{61.6\%}{100\%}\right) = -1.21 \times 10^{-4} \text{ y}^{-1} t$$

$\ln (61.6\%/100\%)$
$= \ln (0.616)$
$= -0.485$

$$-0.485 = -1.21 \times 10^{-4} \text{ y}^{-1} t$$
$$\frac{-0.485}{-1.21 \times 10^{-4} \text{ y}^{-1}} = t = 4.00 \times 10^3 \text{ y}$$

It is reasonable that a loaf of bread made for an ancient pharaoh would be 4000 years old.

Math Review

Introduction

General chemistry classes require lots of basic mathematical skills. These include many which you were taught earlier in your academic career but have probably forgotten from lack of use. In this section we will address those rusty math skills so that you can call upon them as necessary during your study of general chemistry. The important topics to learn from this section are: <u>proper use of scientific notation, basic calculator skills including entering numbers in scientific notation, how to round off numbers, use of the quadratic equation, the Pythagorean theorem, and some rules of logarithms.</u> Most of these skills will be required in various chapters in your general chemistry course.

Math Review Key Equations & Concepts

1. $$x = \frac{-b \pm \sqrt{b^2 - 4ac}}{2a}$$

    This equation is used to determine the solutions to quadratic equations, i.e. equations of the form $ax^2 + bx + c$. You will frequently encounter quadratic equations in equilibrium problems.

2. $$a^2 + b^2 = c^2$$

    The Pythagorean theorem is used to determine the length of one side of a right triangle given the length of the other two sides of the triangle. This formula is frequently used in the section on the structure of solids to determine the edge length or the diagonal length of a cubic unit cell when calculating the atomic or ionic radius of an element.

3. $x = a^y$ then $y = \log_a x$

    $$\log(x \cdot y) = \log x + \log y$$

    $$\log\left(\frac{x}{y}\right) = \log x - \log y$$

    $$\log(x^n) = n \log x$$

    The first equation is the definition of logarithms. The other equations are basic rules of algebra using logarithms. These rules apply to logarithms of any base, including base e or natural logarithms, ln. These equations will be used frequently in kinetics and thermodynamic expressions.

Scientific and Engineering Notation

In the physical and biological sciences it is frequently necessary to write numbers that are extremely large or small. It is not unusual for these numbers to have 20 or more digits beyond the decimal point. For the sake of simplicity and to save space when writing, a compact or shorthand method of writing these numbers must be employed. There are two possible but equivalent methods called either scientific or engineering notation. In both methods the insignificant digits that are placeholders between the decimal place and the significant figures are expressed as powers of ten. Significant digits are then multiplied by the appropriate powers of ten to give a number that is both mathematically correct and indicative of the correct number of significant figures to use in the problem. To be strictly correct, the significant figures should be between 1.000 and 9.999, however this particular rule is frequently ignored and, in fact, must be ignored when adding numbers in scientific notation that have different powers of ten. The only difference between scientific and engineering notation is how the powers of ten are written. Scientific notation uses the symbolism x 10^y whereas engineering notation uses the symbolism Ey. Engineering notation is frequently used in calculators and computers.

INSIGHT:

> *Positive powers of ten* indicate that the decimal place has been *moved to the left that number of spaces*.
> *Negative powers of ten* indicate that the decimal place has been *moved to the right that number of spaces*.

A few examples of both scientific and engineering notation are given in this table.

Number	Scientific Notation	Engineering Notation
10,000	1×10^4	1E4
100	1×10^2	1E2
1	1×10^0	1E0
0.01	1×10^{-2}	1E-2
0.000001	1×10^{-6}	1E-6
23,560	2.356×10^4	2.356E4
0.0000965	9.65×10^{-5}	9.65E-5

It is important for your success in chemistry that you understand how to use both of these methods of expressing very large or small numbers. Familiarize yourself with both methods.

Basic Calculator Skills

General Chemistry courses require calculations that are frequently performed on calculators. You do not need to purchase an expensive calculator for your course. Rather you need a calculator that has some basic function keys. Common important functions to look for on a scientific calculator are log and ln, antilogs or 10^x and e^x, ability to enter numbers in scientific or engineering notation, x^2, $1/x$, $\sqrt{}$ or multiple roots, like a cube or higher root. More important than having an expensive calculator is knowing how to use your calculator. It is strongly recommended that you study the manual that comes with your calculator and learn the basic skills of entering numbers and understanding the answers that your calculator provides. For a typical general chemistry course there are three important calculator skills that you must be proficient with.

  1) Entering Numbers in Scientific Notation

Get your calculator and enter this number into it, 2.54×10^5. The correct sequence of strokes is press 2, press the. button, press 5, press 4, and *then press either EE, EX, EXP or the appropriate exponential button on your calculator*. **Do not press x 10 before you press the exponential button!** This is a very common mistake and will cause your answer to be 10 times too large.

After you have entered 2.54×10^5 into your calculator, press the Enter or = button and look at the number display. If it displays 2.54E6 or 2.54×10^6, you have mistakenly entered the number. Correct your number entering method early in the course before it becomes a bad habit causing you to miss many problems.

2) <u>Taking Roots of Numbers and Entering Powers</u>

Frequently we must take a square or cube root of a number to determine the correct answer to a problem. Most calculators have a square root button, $\sqrt{}$. To take square roots just enter the number into your calculator and press the $\sqrt{}$ button to get your answer. For example, take the square root of 72, your answer should be 8.49. Some, but not all, calculators have a $\sqrt[3]{}$ button as well. If your calculator does not have a $\sqrt[3]{}$ button then you can use the y^x button to achieve the same result. To take a cube root, enter 1/3 or 0.333 as the power and the calculator will take a cube root for you. For example, enter $27^{0.333}$ into your calculator. You should get 3.00 as the correct answer. If you need a fourth root, enter ¼ which is 0.25 as the power, and so forth for higher roots.

3) <u>Taking base 10 logs and natural or naperian logs, ln</u>

Many of the functions in thermodynamics, equilibrium, and kinetics require the use of logarithms. All scientific calculators have log and ln buttons. To use them simply enter your number and press the button. For example, the log 1000 = 3.00 and the ln of 1000 = 6.91.

INSIGHT: A common mistake that students frequently make is taking the ln when the log is needed and vice versa. Be careful which logarithm you are calculating for the problem.

Rounding of Numbers

When determining the correct number of significant figures for a problem it is frequently necessary to round off an answer when truncating to the appropriate number of significant figures. Basically, if the number immediately after the last significant figure is a 4 or lower, round down. If it is a 6 or higher, round up. The confusion arrives when the determining number is a 5. If the following number is a 5 followed by a number greater than zero, round the number up. If the number after the 5 is a zero, then the textbook used in your course will have a rule based upon whether the following number is odd or even. You should use that rule to be consistent with your instructor. The following examples illustrate these ideas. In each case the final answer will contain three significant figures.

Initial Number	Rounded Number
3.67492	3.67
3.67623	3.68
3.67510	3.68
3.67502	Use your textbook rule.

Use of the Quadratic Equation

Equilibrium problems frequently require solutions of equations of the form $ax^2 + bx + c$. These are quadratic equations and the two solutions can always be determined using this formula.

$$x = \frac{-b \pm \sqrt{b^2 - 4ac}}{2a}$$

For example, if the quadratic equation to be solved is $3x^2 + 12x - 6$, then a = 3, b = 12, and c = -6. The two solutions can be found in this fashion.

$$x = \frac{-b \pm \sqrt{b^2 - 4ac}}{2a}$$

$$x = \frac{-12 \pm \sqrt{12^2 - 4(3)(-6)}}{2(3)}$$

$$x = \frac{-12 \pm \sqrt{144 + 72}}{2(3)}$$

$$x = \frac{-12 \pm \sqrt{216}}{6}$$

$$x = \frac{-12 \pm 14.7}{6} = \frac{2.7}{6} \text{ and } \frac{-26.7}{6}$$

$$x = 0.45 \text{ and } -4.45$$

INSIGHT:

> Quadratic equations always have two solutions. In equilibrium problems one of the solutions will not make physical sense. For example, it will give a negative concentration for the solutions or produce a concentration that is outside the possible ranges of solution concentrations. It is your responsibility as a student to choose the correct solution based on your knowledge of the problem.

The Pythagorean Theorem

In Module 13 we determined the radius of an atom in a cubic unit cell. Because the cell is cubic, a right triangle can always be formed using two of the sides and the face diagonal. The length of the face diagonal can be determined using the Pythagorean theorem. An example of the unit cell geometry and determining the face diagonal length is given below.

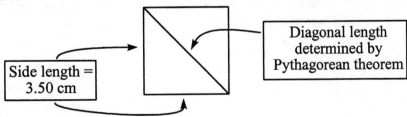

Side length = 3.50 cm

Diagonal length determined by Pythagorean theorem

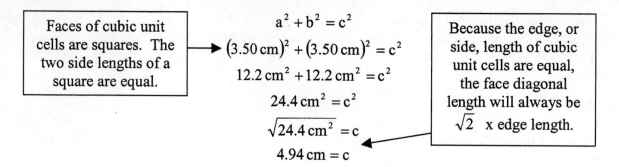

| Faces of cubic unit cells are squares. The two side lengths of a square are equal. | $\rightarrow$ | $a^2 + b^2 = c^2$
 $(3.50\,\text{cm})^2 + (3.50\,\text{cm})^2 = c^2$
 $12.2\,\text{cm}^2 + 12.2\,\text{cm}^2 = c^2$
 $24.4\,\text{cm}^2 = c^2$
 $\sqrt{24.4\,\text{cm}^2} = c$
 $4.94\,\text{cm} = c$ | Because the edge, or side, length of cubic unit cells are equal, the face diagonal length will always be $\sqrt{2}$ x edge length. |

Rules of Logarithms

Logarithms are convenient methods of writing numbers that are exceptionally large or small and expressing functions that are exponential. They also have the convenience factor of making the multiplication and division of numbers written in scientific notation especially easy because in logarithmic form addition and subtraction of the numbers is all that is required. By definition, a logarithm is the number that the base must be raised to in order to produce the original number. For example, if the number we are working with is 1000 then 10, the base, must be cubed, raised to the 3rd power, to reproduce it. Mathematically, we are stating that $1000 = 10^3$, so the log (1000) = 3. There are three commonly used rules of logarithms that you must know. They are given below.

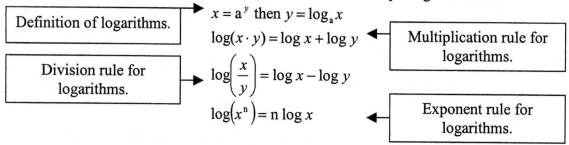

Definition of logarithms.	$\rightarrow$	$x = a^y$ then $y = \log_a x$	
		$\log(x \cdot y) = \log x + \log y$	$\leftarrow$ Multiplication rule for logarithms.
Division rule for logarithms.	$\rightarrow$	$\log\left(\dfrac{x}{y}\right) = \log x - \log y$	
		$\log(x^n) = n \log x$	$\leftarrow$ Exponent rule for logarithms.

INSIGHT: | These rules are correct for base 10, natural, or any other base logarithms.

Significant Figures for Logarithms

When we write this number, 2.345 x 10^{12}, we understand that there are 4 significant figures (the 2, 3, 4, and 5) but the power of 10 (the number 12) is not counted as significant. If we take the log of 2.345 x 10^{12} the number of significant figures must remain the same, 4. The log of 2.345 x 10^{12} = 12.3701... What numbers indicate the exponents that are present in scientific notation? In logarithms, the numbers to the left of the decimal place, called the characteristic, are insignificant and the ones to the right, the mantissa, are significant. Thus the log (2.345 x 10^{12}) = 12.3701 and both numbers have 4 significant figures.

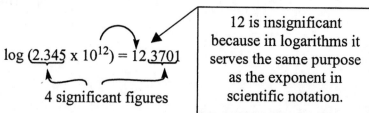

$\log(\underline{2.345} \times 10^{12}) = 12.\underline{3701}$

4 significant figures

| 12 is insignificant because in logarithms it serves the same purpose as the exponent in scientific notation. |

118